Frederick Tracy

The Psychology of Childhood

Frederick Tracy

The Psychology of Childhood

ISBN/EAN: 9783337367343

Printed in Europe, USA, Canada, Australia, Japan

Cover: Foto ©berggeist007 / pixelio.de

More available books at **www.hansebooks.com**

THE
PSYCHOLOGY OF CHILDHOOD

BY

FREDERICK TRACY, B.A., Ph.D.
LECTURER IN PHILOSOPHY IN THE UNIVERSITY OF TORONTO

SECOND EDITION

BOSTON, U.S.A.
D. C. HEATH & CO., PUBLISHERS
1894

INTRODUCTION TO THE FIRST EDITION.

The author has here undertaken to present as concisely, yet as completely, as possible, the results of the systematic study of children up to date, and has included everything of importance that could be found. This work was greatly needed, and has been done with a thoroughness which all interested in the subject will gratefully recognize. Most observations have been limited to one or more aspects of the vast, many-sided topic. As we are now able to catch a glimpse for the first time of the entire field, we realize the importance of results already achieved, and the yet greater promise of the future. The questions here treated are fundamental for both psychology and pedagogy, for the more fundamental the traits, the earlier they unfold. Yet it should be remembered that the data for infant study are relatively more complete than are the records of children of school age. The latter, when they are fully presented, may be more practical, but the former are more fundamental for philosophy and ethics.

It is a most auspicious fact for philosophy and for education, that both are coming to be based more and more upon the eternal and natural foundation of sympathetic observation of childhood, and that the same season that witnessed the completion of this memoir has witnessed the formation of a national society for child study, inaugurated by a successful three days' congress.

This dissertation is far more than a compilation. It brings important additions to our knowledge upon some of the most important topics. This is perhaps most noticeable in the case of the chapter on language, almost a monograph in itself, and which will interest philologists as well as psychologists and teachers.

G. STANLEY HALL.

CLARK UNIVERSITY, September, 1893.

PREFACE TO THE SECOND EDITION.

IN the very short time which has elapsed since the publication of the first edition of this little book, the author has found neither leisure nor material for any very extensive enlargement or modification. A few typographical errors have been corrected, one or two passages slightly modified in conformity with the results of later investigation, and a footnote or reference added here and there. The bibliography has been brought up to date by the addition of several works which have appeared since the publication of the first edition, and at the suggestion of several reviewers, an index and a table of contents have been added. On the other hand, a good many of the reference numbers which disfigured the pages of the first edition, are in the present edition omitted or simplified.

F. T.

UNIVERSITY OF TORONTO, September, 1894.

TABLE OF CONTENTS.

	PAGE
INTRODUCTION TO THE FIRST EDITION, BY PRESIDENT HALL	iii
PREFACE TO THE SECOND EDITION	v
PRELIMINARY	ix

CHAPTER I. SENSATION.

SECTION		
I.	SIGHT	2
II.	HEARING	20
III.	TOUCH	27
IV.	TASTE	31
V.	SMELL	34
VI.	TEMPERATURE	37
VII.	ORGANIC SENSATIONS	38
VIII.	MUSCULAR FEELINGS	40

CHAPTER II. EMOTION.

I.	FEAR	44
II.	ANGER	47
III.	SURPRISE, ASTONISHMENT, CURIOSITY	49
IV.	ÆSTHETIC FEELINGS	53
V.	LOVE, SYMPATHY, JEALOUSY, ETC.	55

CHAPTER III. INTELLECT.

SECTION		PAGE
I.	Perception	62
II.	Memory	65
III.	Association	69
IV.	Imagination	72
V.	The Discursive Processes	75
VI.	The Idea of Self	82

CHAPTER IV. VOLITION.

I.	Impulsive Movements	93
II.	Reflex Movements	94
III.	Instinctive Movements	98
IV.	Ideational Movements	102

CHAPTER V. LANGUAGE.

I.	Heredity vs. Education in Language	115
II.	The Physiological Development	119
III.	The Phonetic and Psychic Development	124

Vocabularies	140
Unpublished Sources of Information	161
Published Sources of Information	162
Index	169

PRELIMINARY.

The comparative method of study has commended itself to all the sciences in modern times by its fertility in results, and is now being employed extensively in two principal directions: viz., the analogical and the genetical. The philologist, for example, compares his own language, on the one hand with other languages (in the search for analogies), and on the other avails himself of all manuscripts, inscriptions, etc., which show him his language in its earliest stages, and help him to determine by the operation of what causes, and according to what laws, it has developed from its original crude and inefficient state to its present polished and complicated condition. And similarly with other sciences. In the case of psychology the application of the comparative method has led the investigator to the observation of mental manifestations in the lower animals; in human beings of morbid or defective mental life, such as the insane, the idiotic, the blind, deaf and dumb; in peoples of different types of culture, ancient and modern, savage and civilized; and finally to the study of mental phenomena in their genesis and early development in the life of the child. If the child is only the adult in miniature, and if society is only the individual "writ large," then in studying the infant mind we are approaching a vantage ground from which we may catch a prophetic view, not only of psychological, but also of sociological phenomena.

When we compare the young child with the young animal, we cannot fail to be struck by the apparent superiority of the latter over the former, at the beginning of life. The human infant, for example, requires weeks to attain the power of holding his head in equilibrium, while the young chicken runs about and picks up grains of wheat before the first day of his life is over. This, however, carefully considered, is a token rather of the superiority than of the inferiority of the human being. The higher you ascend in the scale of being, the more varied and complex is the environment in which the individual moves, and to which he must adapt his movements. This adaptation requires, on the physiological side, a cerebral and nervous development, and on the psychic side a mental growth, for which *time* is an absolute necessity. Animals go on all their lives, doing the same simple things, which require a minimum of mental activity, and which, by dint of constant repetition, produce physiological adjustments that become at length hereditary; so that phenomena which seem to the casual observer the index of an astonishing degree of mental advancement — such as the "scampering" of young chicks on a certain peculiar call of the mother — are really at bottom little more than the response of an organism, adjusted by heredity, to the action of an external stimulus.

The longer and more arduous the journey, the more time

is required for preparation; the more complicated the art to be acquired, the more extended is the period of apprenticeship. So the child, having an infinitely grander life before him, and infinitely more exalted, complicated and difficult operations to perform — mental, moral and physical — requires a longer period of tutelage than the chicken, which on the first day of his life scratches and pecks, and to the end of his existence makes no advance upon these simple operations. The young animal, before the end of the first day of his life, does what it takes the child a year to accomplish; but the child of two years does what the animal never will accomplish to the end of his days.[1]

The object of the present essay is to discuss infant psychology. When and how do mental phenomena take their rise in the infant consciousness? How far are they conditioned by heredity, and how far by education, including suggestion? What is the nature of the process by which the automatic and mechanical pass over into the conscious and voluntary? These are some of the questions to which the following pages may help to furnish an answer. That they may do so, it has been thought best to gather together, so far as possible, the best work that has been done in actual

---

[1] "Es scheint ein Naturgesetz zu walten, dass das höhere Bedeutende sich langsamer entwickele, und sich durch die langsamere Entwickelung eine längere Dauer gleichsam erkaufe." Sigismund: "Kind und Welt," p. 17. See also on this subject, Jastrow: " Problems of Comparative Psychology," *Pop. Sci. Mo.*, Nov. 1892. It should also be noted in this connection that the intra-uterine period is relatively much shorter in man than in most of the lower animals. The horse, for example, lives a much shorter life than man, and yet his preparatory fœtal stage is actually longer.

observation of children up to the present time, arrange this under appropriate headings, incorporate the results of several observations made by the writer himself, and present the whole in epitomized form, with copious references and quotations. The inquiry proceeds along the line usually followed by psychologists, and treats the mental endowment, from the genetic point of view, in the following order: sensation, emotion, intellect, volition; child-language, on account of its paramount importance, being treated in a chapter by itself. It was intended at first to add a chapter on the moral nature of the child, but as the work progressed, it became more and more evident that, to treat this important phase of child-life adequately, would require not only more space than is at our disposal at present, but an advance into later stages of life than are embraced in the present work, which is intended only as a manual of *infant* psychology in an approximately strict sense of the words.

I cannot forbear calling attention in this place to one great general principle, which is so constantly illustrated in the child's mental life that it may be considered universal. It might be appropriately named the principle of transformation, and explained as follows: Every mental phenomenon passes through a graduated ascending series of development. At first, the physiological predominates, consciousness is at a minimum, and the so-called *mental* phenomenon would be more accurately defined as the reaction of the nervous system to external stimuli or to organic conditions. For example, the child *cries* at intervals from the moment of his birth, but at first this cry is independent of his will, and possesses scarcely any mental significance, for it is made without cerebral coöperation, and — as in the case of microcephalic infants — even when the cerebrum is entirely absent[72].[1] Later the mental aspect becomes more prominent. When the intellect and will have become sufficiently developed, the child directs his attention to the act, makes it his own and performs it voluntarily. The process perhaps has not changed at all, to outward appearance, but when viewed on the inner side, it is seen to have been completely *transformed* in character; and one of the most difficult tasks for the psychologist is to determine the *when* and the *how* of this transformation.

The exact *time* at which each psychic activity makes its appearance, is perhaps of less importance than the *order* of the various activities; yet in order to ascertain the latter, the former must be carefully attended to. Hence both absolute and relative times receive considerable attention in the following pages.

---

[1] The numbers in brackets are references to the bibliography at the back.

THE PSYCHOLOGY OF CHILDHOOD.

CHAPTER I.

SENSATION.

It is important to treat sensation first, because it lies at the foundation of all mental development. All the higher processes of mind are simply the result of progressive "syntheses of the manifold" as given in sensation. Though we may not agree with Locke, that all ideas are derived *from* sensation, yet we must agree that there are no ideas in the mind *prior to sensation*. And looking at the active side of our nature, the intimate connection between the senses and the will is equally manifest. Our sense-impressions, produced by external objects upon the peripheral organism, are conveyed along the afferent nerves to sensory centres closely connected with corresponding motor centres in the cerebral cortex. Hence the importance of the child's sense-growth.

Are any sensations felt in the fœtal stage of existence? And if so, what? In answer to this question, we may, first of all, proceed negatively and determine those senses which obviously cannot be in operation at this time. Any sense requiring as the condition of its exercise the medium of light or air, cannot operate until the child is born, for prior to this time he does not come into contact with these media. On this ground, sight, hearing and smell are probably to be excluded; the first on account of the darkness of the uterus, the others because the auditory and nasal passages are at this time entirely filled with the amniotic liquid, to the exclusion of all air, even if this were available. There is reason to believe, however, that from about the middle of this period the fœtus is susceptible to changes of temperature[52], and that touch is to some degree awakened by contact with the surrounding matrix[47]. To what extent these rudimentary fœtal sensations partake of the truly psychic character is of course very difficult to determine. Many psychologists[1] are of the opinion that they do not at all involve the coöperation of the centres of sensational and motor ideality. Nevertheless, it is certain that during the later months of pregnancy, very great changes take place in the embryonic brain, especially in the cerebrum.[2] If it be allowable to conjecture, it is probable that the "sensations" of the embryo involve consciousness, though very dim and vague, and that the fœtal movements are reflex or automatic, taking place in virtue of an organic connection between feeling and movement, due in large part to heredity.

I. Sight.

The Embryonic Eye.—During the earlier stages of the embryonic growth, the head is much larger in proportion to the other parts of the body than at any subsequent time; and this is especially noticeable in the anterior regions, where the primary vesicle bulges out prominently on each side. These protruding portions gradually fold in upon themselves to form the nervous parts of the eye, such as the retina and optic nerve. Simultaneously with this,

---

[1] E.g., Wirchow, quoted by Perez.  [2] Bastian.

the crystalline lens is developed by the involution of the epiblast, and is received into the hollow cup formed by the folding in of the primary vesicle spoken of. The remaining space afterwards becomes filled with the vitreous humor. "The lids make their appearance gradually as folds of integument, subsequently to the formation of the globe in the third month of fœtal life. When they have met together in front of the eye, their edges become closely glued together by an epithelial exudation which is removed a short time before birth."

We have already remarked that no sensations of sight are received during the fœtal period. If this be true, the cause lies, not in the imperfection of the organ itself — for the experiments of Kussmaul and Genzmer on premature children, show that at least two months before the normal birth-time, the mechanism of the eye is fully developed and capable of reaction to appropriate stimuli — but in the absence of light-impressions. There may even be at this time vague sensations of light, arising from subjective or intrauterine causes, though if there be, they can have but little psychological importance, and can by no means account for the actual functioning of the eye immediately after birth.

THE EYE OF THE NEW-BORN. — If, therefore, the statement is made that the new-born child is blind, it must not be taken to mean that he is in darkness — for the peripheral mechanism of the eye is complete at birth, and the difference between light and darkness is felt from the beginning — but only this, that he cannot as yet *see things*, in the proper sense of the terms. This is due to lack of experience, to imperfect development of the cerebral centres, and to the dazzling effect of the light, which now streams in, as Sigismund says, with millions of waves, upon a delicate organ, accustomed, up to this time, to complete darkness.¹ This latter obstacle, however, is soon overcome, and the child's progress in seeing takes place with great rapidity.

The sensation of *light* is the first feeling, having an external cause, which the child experiences by means of the eye. This organ is especially adapted, by its peculiar mechanism of retina and rods and cones, and by its nerves and muscles of convergence, contraction and accommodation, to receive the rays of light that fall upon it; and hence, as soon as the first shock is over, and the infant eye has become accustomed to its new surroundings, it turns toward the light as naturally as the opening petals of a newly-blown flower turn toward the rising sun. Or, as Locke has said: "Even as the soul thirsts for ideas, so the eye of the child thirsts for the light." This sensibility to light is normally present in the first minutes of life, and is rarely delayed beyond a few hours, except in the case of some malformation of the organs[72]. At this stage, however, the distinction of light and darkness is *felt* rather than *known;* and even the turning of the head toward the light, which has been observed on the second day of life, and even as early as the twentieth hour[110],² must be considered as nearly akin to the movement of the plant toward the light. But this condition of things is not of long duration. To take a single case (that of Preyer's boy), we are told that he soon began to show signs of pleasures at a moderate light, pain at too powerful glare, and less pleasure in dark-

---

¹ Kussmaul also remarks: "Ausgetragene Kinder, welche eben zur Welt gekommen und ruhig geworden sind, versuchen öfter das Auge wiederholt zu öffnen sind aber immer wieder gezwungen es rasch und kramphaft vor dem einfallenden hellen Lichte zu schliessen."

² Kussmaul cites the case of a boy, who though born in the seventh month, yet turned his head towards the window on the second day of his life.

ness. Even during the first day the expression of his face changed when an intervening object cut off the light, and on the eleventh day he would cry when the light was carried out of the room. As time passed on, he continually took increasing notice of these sensations, until in his second month the sight of a bright light, or a brightly colored object was sufficient to elicit from him exclamations of delight.

Too powerful a light causes discomfort, even in sleep. The child knits his eyelids more closely together, or even becomes restless and awakes. A very bright light is especially painful immediately on awakening. Preyer observed that his boy shut his eyes and turned his head away when a candle was held close to him on awakening. But when he had been awake for some hours, he looked steadily, without blinking, at a candle held one metre from his eyes.[1]

With these qualifications, we may conclude, then, that "light is pleasant to the eye," being its natural "food," and that under its influence the delicate organ of vision grows and develops, the visual centres in the cerebrum become differentiated and capable of performing their function, thus rendering possible the subsequent apprehension of qualities in external things by means of this sense.

PHYSIOLOGICAL ADJUSTMENTS TO LIGHT. — At the beginning of life, all adjustments of the visual organ to the strength of the light are reflex. For example, from the very first the filaments that contract the pupil perform their function. The pupil accommodates itself to the brightness of the light, expanding and contracting, as Kussmaul and Raehlmann have shown. Both pupils contract when the light reaches one of them. These movements of contraction

---

[1] I believe this sensitiveness to light on first awaking is also quite common among adults.

remain automatic to the end of life. It is otherwise with such movements as following a moving light or object with the eyes. This is at first undoubtedly reflex, since it takes place before the conscious centres have been sufficiently developed for voluntary action, but it afterwards certainly comes within the domain of the will, as is evident from adult conscious experience.

Eye Movements. — This includes movements of the eyeballs (upward, downward, and from right to left, etc.), and movements of the lids (raising and lowering), as well as the relation of the two to each other.

Does the child possess a complete nerve-mechanism for eye-movements, working perfectly from the beginning, or does he gradually and painfully acquire all eye-movements? The most recent observations lead to the following conclusion: The mechanism is inherited complete so far as pupil, retina and nerve tracts are concerned, but the corresponding brain centres are not yet developed in the first days, and become so only by experience; consequently the adjustment of movements to external conditions takes place by degrees. No doubt there is a hereditary predisposition to coördinated movements, which to some extent facilitates the subsequent adjustment, but the largest share is due to experience. The following facts have been established by careful observations:

First. — As to movements of the eye-balls: Complete conscious coördination of the movements of the two eyes does not take place during the first days. True, the eyes sometimes move together, even from the first,[1] but there are also numberless non-coördinated movements, which proves that

---

[1] According to one observer on the fourth day, according to another on the second day (B), while a third noticed them five minutes after birth[72]. Miss Shinn found these movements usually symmetrical from the first[87].

the coördinated ones are accidental at first, and that the useless movements are only gradually eliminated. Raehlmann and Witkowski, in a very large number of observations on new-born children, carried on for fifteen years, found that the infant eyes, especially in sleep, "assume positions and perform movements which are entirely contrary to all the principles of association," including complete opposite movements of the eyes, resulting in divergence of eye-positions[78]. Sometimes the eyes move together, laterally and vertically (though this coördination is not so perfect as in the adult), but just as frequently are the movements irregular (E). Sometimes one eye moves, while the other remains at rest. Sometimes the head is turned in one direction, and the eyes in another. A great deal of unnecessary convergence takes place, as I have frequently observed. In most observed cases, however, these asymmetrical movements have become very much less frequent by the third month, and, at a little later time, have almost entirely disappeared, except in sleep.

Second. — As to movements of the lids: The only lid-movement that can be accepted as inborn, is the sudden "blinking" when a foreign substance comes into contact with the lashes or the cornea, or on the sudden approach of a strong light. The mere *approach* of the object, without contact, does not produce blinking at first; indeed, in some cases, it fails in children two months old[78]. All other lid-movements are at first accidental. Sometimes the lids move together, though more frequently they do not. Sometimes one eye remains open while the other is shut. The two eyes do not always open to an equal degree; and often, if one eye be disturbed and blinking take place, the lid of the undisturbed eye will follow some time after the other. The lids are often raised while the look is directed downward, and *vice versa*. The child often falls asleep with the lids a little apart. ⁻Coördination, then, is not perfect at first, but becomes so by experience. Not only so, but the child actually has to *unlearn* several movements (*e.g.*, raising the lids while the eyes are directed downward) and these have become impossible in the adult[15]. Gradually these asymmetrical movements disappear, until by the end of the third month they have become very rare, except in sleep.

All that has been said concerning movements of the eyes, and of the lids, separately, is true, *mutatis mutandis*, of the relation of these to each other. Perfect coördination among the several branches of the oculomotorius is not present at the beginning of life (not at all during the first ten days, according to Raehlmann), but is a gradual attainment, requiring time and experience. But when once the awakening mind has taken possession of the eye, and made the movements of that organ its own, it becomes one of the most expressive organs of the body, and reveals the various shades of the inner feeling with astonishing accuracy.

FIXATION. — By this is meant conscious direction of the gaze upon an object, as contrasted with passive staring into space. And the question of most importance here is: When does the child pass from the one to the other? The question is important, because it throws light upon the beginning of volition, which, in its exercise, determines in such large measure the mental and moral development of the child.

Preyer divides the "seeing" of the infant into four stages. I shall follow his classification, bringing under each heading also the observations made by others on the period in question:

First. — Staring into empty space; experiencing a sensation, but not perceiving an object. The ability to "fixate" an object is lacking in the newly-born, because he has as yet no control over the muscles that move the head and eyes. The apparent looking of the first days is not, therefore, a voluntary or intelligent action, but only the instinctive turning of the head and eye so as to bring the light into contact with the central portion of the retina, where it produces the greatest amount of pleasurable feeling. When Champneys observes that one child "fixed" his eyes on a candle on the seventh day, and Darwin reports that another child did the same on the ninth day, Preyer remarks that this was probably not real looking, but only staring into space, since in other similar cases it was observed that the child continued to "look" when the object was withdrawn. There is probably no fixation in the first nine days.

Second. — The child no longer "stares," but "looks." He fastens his gaze upon a bright extended surface (*e.g.*, his mother's face) and when another bright, moderately large object comes within the field of vision, he turns his eyes from the first to the second. One child was observed to do this on his eleventh, and another on his fourteenth day. Along with the fixing of the gaze, there is also a more intelligent expression. Perez reports that a child observed by him "looked fixedly for three or four minutes at a flickering reflection of light before the end of his first month." In another case, an object was looked at steadily in the fourth week for the first time; in another, a yellow dress held the child's gaze at five weeks; and in still another the power of fixation is reported on as still absent when the child was two months old (E). Sigismund observes that about the middle of the first three months the child "begins to look at objects with attention;" and Raehlmann found that "appropriate selection among the many possible eye and lid movements, with fixation of the object, took place for the first time after the fifth week."[1]

---

[1] Taking the average of the above cases, we have the thirty-second day, or during the fifth week, as the time of the beginning of fixation.

Third. — In the third stage, the child has acquired the power to follow with his eyes a bright, moving object. Here we have associated movements of the eyes, the head being motionless, or nearly so. We have now, therefore, a distinct advance, requiring a higher exercise of power over the muscles. The movement is not accomplished if the object be moved too rapidly. In one case the child's eyes followed a moving candle in the second week; in another, on the twenty-third day. But most of the observers have noticed this activity first about the fifth week, some as late as the sixth or seventh. Raehlmann remarks on this point to the following effect: Associated lateral movements of the eyes can be found seldom earlier than the fifth week. Hold a bright or colored object at a little distance, directly before the child's eyes. One soon notices a peculiar change of expression, accompanied by cessation of the movements which the limbs until now were executing. The object has been fixated. Now move it slowly in a horizontal direction to one side, and both the eyes follow, but without movement of the head. If the object be moved quickly, the child's eyes lose it at once; and also if the movement be vertical instead of horizontal.[1]

In the early part of this third stage, Preyer holds, there is no necessary coöperation of the cerebrum, but only of the corpora quadrigemina, and he cites in proof the experiment of Longet with a pigeon, from which the cerebral hemispheres had been carefully removed, and which, in that condition, followed with its eyes the flame of a moving candle. It may be remarked, however, that since the instinctive and reflex play so much larger a part relatively

---

Genzmer, on the other hand, by shaking a bright object before the eyes, obtained not only fixation, but "following" movements in a large number of children, at a much earlier age than this.

in the lower animals than in man, this proof is not entirely trustworthy, forasmuch as a movement, which in the lower animals is reflex, may in man require the coöperation of the cerebrum. More to the purpose would be the case of an acephalous or microcephalous child. Kollman says of the microcephalous Margaret Becker, eight years of age: "Her gait is tottering, the movements of the head and extremities jerky, not always coördinated, hence unsteady, inappropriate, and spasmodic; *her look is restless, objects are not definitely fixated.*" This case seems to point in the opposite direction from that of Longet's pigeon, and Preyer's conclusion therefrom.

Fourth. — Here we pass from looking to observing, to the active search for objects. The child has acquired ability to give definite direction to the gaze, and hold it there. Of course the first attempts are often ineffectual, but, roughly speaking, from about the third to the fifth month, this power is obtained[78]. A girl of ten weeks looked for the face of a person calling her. A boy in his sixth week moved his head to follow a look cast in a certain direction[66]. Another began in his sixteenth week to look intently at his own hands. Another of twelve weeks, on hearing a noise made by a person on a drinking glass with a moistened finger, turned his head in the direction of the noise, and, after one or two ineffectual attempts, found the object with his eyes and fixated it. In the fourteenth week he followed promptly the movements of a pendulum which made forty complete oscillations per minute[72]. Sigismund's boy, at nineteen weeks, paid great attention to the movements of a pendulum, and afterwards followed the movements of a spoon from dish to mouth and back again, with eager mien. Rapid movements, however, are not as yet preferred. In the railway carriage, the child of this age does not look at the passing objects, but rather at the walls and ceiling of the coach. Not before the twenty-ninth week (in one observed case) did the child look distinctly, beyond doubt, at a sparrow flying by. Another "watched the flight of birds" when five months old [31]. It will readily be observed that the full attainment of this fourth stage involves voluntary control of the mechanism of the eye as well as considerable progress in the intellectual apprehension of the external world. So that now the child is no longer the reflex, staring creature, but has become the *bona fide* "seeing" human being.

SEEING IN PERSPECTIVE. — Numerous observations confirm the following statements:

(*a*) The new-born child does not see, in any sense of the word, objects that are very distant from him; or if he sees them at all, the impression made by them upon the retina is so vague as not to enter into distinct consciousness. Indeed, there are few distinct retinal images at first from objects either near or distant.

(*b*) For a long time after he is able to see objects at a considerable distance, and several objects at unequal distances in the field of vision together, he still does not know how unequal their distances are, or even that they are unequal.[1] The physiological mechanism of the eye, by which it is "accommodated" to the distance of the object seen, operates very early; but the estimation of distance is long imperfect. At one month and five days, Tiedemann's son "distinguished objects outside him, and tried to seize them, extending his hands and bending his body." By the end of the second month, there is, according to one observer,

---

[1] "Il est prouvé, par des faits certains, qu'ils sont plusieurs mois, sans avoir d'idée précise des distances." Cabanis, "Rapports du physique et du moral de l'homme" (47).

a vague idea of distance. But most observers place it much later than this. One says: "The first real grasping of the fixated object, with appreciation of its distance, was observed about the end of the fifth month. But it is very slowly acquired, and not until much later than this does the hand proceed directly, by the nearest way, to the object"[78]. Another found but little comprehension of size or distance until the sixth month. Another reports of a little boy that when nearly a year old, he "saw the moon and stars, and his eagerness to have the moon was most interesting. Night after night he would call for it, stretching out his little hands towards the window"[11]. The girl F. did not look at anything very far away until she was a year old. Another child, even in the second year, "repeatedly misnamed men or boys at perhaps twenty yards distance; the less familiar person being almost always called by the name of the one better known"[87]. Preyer's boy, when four months old, "often grasped at objects which were twice the length of his arm from him; when considerably over a year old he grasped again and again at a lamp in the ceiling of a railway carriage, and when nearly two years old tried to hand a piece of paper to a person looking out of a second story window, from the garden below — "a convincing proof how little he appreciates distance."[1]

(c) At first the child sees only colored surface, and not figures in the third dimension. All objects present themselves to his eye simply as patches of color. Gradually, by the aid of movement and touch, he comes to a knowledge of their cubic properties. Hence also arises by experience an

---

[1] And yet another child had apparently attained a comparatively correct estimation of distance by the end of her seventh month, as she "invariably refused to reach for an object more than fourteen inches distant, her reaching distance being from nine to ten inches"[6].

association between the forms and distances of objects and their varying degrees of luminosity, so that the child comes to interpret the one in terms of the other. Hence the progress of the child in complete vision, including all that is meant by the appreciation of perspective, is immensely facilitated from the time he begins to walk, since, by locomotion, he is able to approach the object and bring sight, touch, and the muscular sense to bear upon its examination.

COLOR DISCRIMINATION. — Not only is color blindness "notoriously hereditary" as an abnormal condition in the adult [79][1], but it is the normal condition of the new-born child. Since the tractus opticus does not get its nerve medulla, and with that its permanent coloring, until the third or fourth day of life, there is probably no discrimination of colors up to that time, but only of light and darkness. Moreover, even when discrimination of colors has begun, it proceeds very slowly, and the investigation is beset by difficulties. How are we to distinguish (*e.g.*) the mere feeling of difference between sensations of color from intelligent apprehension of the colors themselves? Very little can be done until the child can speak, and even then new difficulties present themselves. The names of colors are more difficult to acquire than the names of things, because more abstract. Grant Allen found that children of two years and even more, who knew perfectly well the names of grapes, strawberries, and oranges, yet had no appropriate verbal symbol for purple, crimson, or orange, as a color [8];

---

[1] Color blindness seems much more common among males than among females. Tests made in 1879 on nearly thirty thousand students of the various schools in the city of Boston, showed that of the boys four in every hundred were color blind, while among the girls the proportion was less than one in a thousand. B. Joy Jeffers, A.M., M.D., in "School Documents," No. 13, Boston, 1880.

and I have found in examining the child-vocabularies, which I have collected for the fifth chapter of the present work, that out of five thousand four hundred words, only about thirty are color terms. In several cases the vocabulary of a child two years old contains not a single color word, though he habitually employs from three to five hundred words [105]. Another difficulty lies in the association between the color and its name. The child may know a color — red — perfectly well; and may also know the sound — red, — but he may not be able to associate the two together, so as when red is named, to point it out; or, when it is pointed out, to name it. This is not from lack of ability to distinguish color from color, but from inability to associate the color with the spoken word.

A girl ten days old had her attention arrested by the contrasted colors of her mother's dress. She seemed pleased and smiled [100]. A boy twenty-three days old was pleased with a brightly colored curtain. Another child in his second month took notice of the difference between bright colors and quiet ones, and showed his preference for the former by smiles. Another, towards the end of his second month, was attracted by white, blue and violet, other colors being indifferent. A girl of three months and a boy of five months seemed pleased with some drawings of a uniformly gray color [66], while Genzmer's boy for the first four months of his life seemed attracted only by white objects, but after that time he began to show a preference for other bright colors, especially red. Raehlmann found no distinction of similar objects differently colored until a good while after the fifth week. Sometimes a strange antipathy to certain colors is manifested. In several cases children have refused to go to anybody dressed in black.

Experiments in color discrimination, which involve the use of words, may be carried on in two ways. A color may be named, and the child required to pick that color out of several; or the color may be shown him, and he required to name it. Preyer used both methods, with the following results: In the twentieth month repeated trials yielded absolutely no result, but in the beginning of the child's third year, the first correct responses were obtained, the result being eleven right answers and six wrong ones. In this case he used two colors, red and green. Then yellow was added, and at once took its place as the color most readily perceived (26th month). The percentages of right answers were: Yellow 82, green 77, red 72. Blue was then added, with the following result: Yellow 94, green 79, red 70, blue 69. Trials made a week later with five colors resulted as follows: Yellow 100, violet 92, green 90, red 83, blue 42. Then, with six colors: Yellow 96, violet 95, red 84, gray 83, green 74, blue 67 (26th and 27th months). Finally, two weeks later, trial was made with nine colors, resulting as follows: Yellow, gray, brown, and black 100, red 91, violet 85, green 36, rose 33, blue 23. Preyer carried these experiments a good deal further, and varied the method, but with substantially the same results. The summary of all his tests up to the 34th month gives the following order of preferences: Yellow, brown, red, violet, black, rose, orange, gray, green, blue. When yellow and red were removed, the child showed less interest. Blue and green were avoided, and mostly named wrong, green being often called "garnix" ("gar nichts" = "nothing at all").

Binet[10] made a number of experiments with a little girl from the 32nd to the 40th month, with results which I may epitomize as follows:

1st series: Red 100, green 61, yellow 52.

2nd series: Red 100, blue 92, maroon and rose 89, violet 75, green 71, white 62, yellow 38.

In these experiments, the child was required to point out

the color named to her. The method was now reversed, and the child required to name the color pointed out to her. The result was as follows:

1st series: Red 100, yellow 0.

2nd series: Blue 100, red 96, green 82, rose 57, violet 54, maroon 50, white 45, yellow 28. (M. Binet says every time an error is committed with yellow, it consists in confounding it with green. He noticed also that violet was confounded with blue.)

Some remarkable differences may be noticed between the results of these two observers. For example, in the perception of yellow: while Preyer's child perceived this color better than any other, Binet's little girl had the greatest difficulty with it. Also as regards blue: in the one case this color stands at the very bottom of the list, while in the other it is almost at the top.[1]

The greatest uniformity obtains in the case of bright and glaring colors, such as red.[2] This may have a physiological basis in the fact that when the eyes are closed in a bright light, red is the only color visible.

In the foregoing experiments, the child must know the names of the colors before the tests can be made; and we can never be certain that the mistakes committed do not arise from confusion of words rather than of colors. On this account, the following tests made by Binet seem to me of far greater value. Instead of the "methode d'appellation," as he calls the system just explained, he adopted here

---

[1] Experiments made by Wolfe on the school children of Lincoln, Nebraska, gave results differing from both Preyer and Binet. Following is the order in this case: White, black and red (nearly always correctly named), then blue, yellow, green, pink, orange and violet, in the order named (105).

[2] Though in the case studied by Miss Shinn red gave a good deal of trouble.

the "méthode de reconnaissance," which consists in showing the child a counter of a certain color, then shuffling it together with a number of counters of that color and others, and requiring him to pick out a counter of that color. In this way the name is not used at all, and the test proceeds purely on the recognition of color. The results by this method were much more satisfactory. With three colors — red, green and yellow — no mistakes were made; and even with seven colors, and with an interval of time between the perception and the recognition, the errors were very few indeed. This seems to show that the child's chief difficulty is not in recognition of the color, but in association of the color with the sound of its name.[1]

OBJECTIVE INTERPRETATION. — The understanding of the meaning of the visual sensation is the slowest in development of all the faculties connected with the eye. The subject belongs indeed properly under the head of Perception and Judgment, and little need be said upon it here.

To comprehend the distance and form of an object, is an advance on the rudimentary "seeing" of the object; but to understand *what* the object *is*, so as to distinguish it from other objects, and be conscious of a relation between it and the perceiving subject, constitutes a still further advance. The child attains this further advance slowly and painfully, at the cost of many tumbles and scratches, the result of errors in judgment that are sometimes pitiable, often comical. Feeling and instinct render great service at this time, and often lead the child to do things which, on a casual view, might too readily be interpreted as the work of judg-

---

[1] For a criticism of all these methods, and the explanation of another, in which the whole question is viewed from the motor standpoint, see two articles by Prof. Baldwin, in *Science* for April 21st and 28th, 1893.

ment; as in the case of the child of less than a month, who made a wry face at the sight of some bitter medicine.

The first object to be recognized is usually the mother's face, which is greeted with a smile of pleasure by children only a few weeks old. But this first recognition is very vague and inaccurate, as is shown by the fact that the infant "recognizes" in the same way, at first, any other face which resembles hers in broad outlines; and that when recognition of the father's face takes place, the child bestows his smile of welcome also on any other bearded gentleman who happens to come within his range of vision. For a long time, objects are not grasped as comprehensive wholes, but rather some striking feature is apprehended, and all else left out of account. Hence arise some of the very peculiar association groupings, which we shall notice in connection with language. From about the sixth month, however, evidences of intelligent comprehension of many of the more common objects may be observed. The smile or nod of the parents is distinguished from that of strangers, and responded to in a different manner. Visual impressions connected with food and clothing are quickly and surely recognized [72]. Yet even much later than this, many mistakes are made. The child of a year and a half will try to pick up a sunbeam from the floor, to grasp his own reflection in the mirror, to pull a stream of water flowing from a sponge, as though it were a string. Even at the close of his second year, pictorial representation is a great mystery to him, and he prefers the reality. Sigismund's boy, at two years, called a circle "plate," a square "bonbon," and his father's shadow "papa;" and Preyer's boy, much later than this, called a square "window," a triangle "roof," a circle "ring," and several dots on the paper "little birds." Pollock tells of a girl nearly two years old, who, on seeing a row of dots on a printed page, thus . . . . , cried out, "Oh, pins," and made repeated attempts to pick them out [69]; and the girl F. was observed one day trying to "pick up" her father's white protruding cuff from what she supposed was the underlying coat-sleeve, as she attempted to grasp the cuff from that side, and seemed much surprised at her failure.

II. HEARING.

The importance of hearing as a knowledge-giving sense would be difficult to overestimate. Besides being the channel of a large part of our knowledge, and the medium of a vast amount of refined pleasure, the sense of hearing plays so large a *rôle* in the acquisition of language that a child who is perfectly deaf from birth, does not learn to speak.

THE EMBRYONIC EAR. — According to Quain's Anatomy, the more important parts of the organ of hearing are formed by the involution of the epiblast from the surface of the head, in the region of the medulla oblongata, by which a depression is produced. This depression gradually deepening, and its outer aperture becoming narrowed, a flask-like cavity is formed, which constitutes on each side the primary auditory vesicle.

The possibility of hearing in the intra-uterine stage, depends on two things, viz., the presence of adequate stimuli, and the permeability of those passages and nerve tracts by which sensations of sound are mediated. As to the first condition, there are probably numerous sounds which *might* produce sensations of hearing in the fœtus, such as the visceral movements of the mother and those of the fœtus itself. Hearing at this stage is, however, highly improbable, because the second condition is not fulfilled. The drum cavity is filled with a viscous mass, which probably prevents the passage of the necessary sound-vibrations through the tympanum, even leaving out of account the complete absence of air at this period. The tympanum itself also has not, at this time, the perpendicular position which it afterwards assumes, and which seems necessary for the transmission of sound, but lies rather in a horizontal situation [86].

HEARING IN THE NEW-BORN. — Czerney, in his experiments as to the comparative soundness of sleep at different times, was unable to use a sound stimulus with new-born children as he did with adults, because of their failure to react to sound-impressions; he was obliged, in their case, to resort to electrical stimulation. Kroner assured himself by many experiments that the child, in the first week of his life, reacts distinctly to strong sound-impressions, and the very careful experiments of Moldenhauer confirm this conclusion. Mrs. Talbot says of one child that he was sensible to sound three hours after his birth. Sigismund saw the first evidences of hearing much later.[1] Perez thinks there may be — through vibration — something corresponding to a rudimentary and general sense of hearing in the uterus. Champneys could not elicit any response — by starting or otherwise — during the first week, to any noise, however loud, unless accompanied by vibration other than air-vibration. Kussmaul utterly failed to produce any impression in the first days, no matter how loud or discordant the noise.[2] He believes hearing sleeps most deeply

[1] " Nach einigen (drei bis acht) Wochen sieht man das Kind bei plötzlichem Geräusche zusammenfahren. Da erkennt mann klar, dass jetzt auch für die wahrnehmende Seele, das Hephata! gesprochen ist." " Kind und Welt," p. 27.

[2] " Mann kann vor den Ohren wachender Neugeborner in den ersten Tagen die stärksten disharmonischen Geräusche machen, ohne dass sie davon berührt werden."

of all the senses. But he quotes Herr Feldsbausch, assistant in midwifery at the hospital in Jena, to show that there was hearing in many cases from the third day. Genzmer found that almost all the children on whom he experimented, on the first day, or certainly on the second, reacted to impressions of sound; but the reaction was unequal in different children. Dr. Deneke found one child of six hours who started and closed his eyes tighter at the sound of two metallic covers striking together; while Preyer observed one who did not react at all on the third day, and another who, on the sixth day, reacted only very slightly. Sully noticed, on the second day, a distinct movement of the head in response to sound, and this is confirmed by Professor Baldwin. Burdach declares the child hears nothing during the first week.

On these the following observations are in place, and may help to the understanding of the discrepancies:

(1) There is unanimity on one point: No one has succeeded in proving that any child hears anything during the first hours. This corresponds to the physiological facts that the eustachian tube is not permeable, nor does air find its way into the middle ear until some little time after respiration has begun. Lesser's experiments show that the foetal conditions of the middle ear may indeed persist in the prematurely born more than twenty hours.

(2) Starting in response to a loud noise may often be caused by vibrations which affect the whole body, and act as a nervous shock. Children are known to start on the slamming of a door, when they make no such response to a voice, however loud. No doubt, in the first case, the child *feels the jar* rather than *hears the noise*.

(3) Any further discrepancies not resolved by these two considerations, may be accounted for by the differences in maturity of different children at birth, and the varying

rapidity with which the physiological adjustments are completed. Generalizing, we may say that the period of beginning to hear varies, according to these circumstances, from the sixth hour to the third week. If, in the fourth week, a healthy, normal child makes no response to a loud sound behind him, there is reason to fear that he will be deaf and dumb [72].

As regards *localization of sounds*, the ear does not render very much service in this, on account of its comparative immobility. Even in the adult, a sound made in the room above is with great difficulty distinguished from a sound made in the room below, unless some other circumstance enter in to assist in the determination.

Champneys' child, on the fourteenth day, turned his head in the direction of his mother's voice, but this was probably due as much to feeling her breath upon his cheek as to hearing, since he did not do it when her face was turned in another direction. Leaving this observation, then, out of account, I find that the period in which children are first observed to turn the head in the direction of sounds, extends from the tenth week [99] (or the fifth week, according to Alcott) to the seventeenth week [21]. One child sometimes turned towards a sound in the sixteenth week.[1] Another, at four months and ten days, "always turned his head exactly in the right direction" [101]. A third turned his head towards a sound for the first time in the eleventh week, and by the sixteenth week this movement had assumed all the certainty of a reflex [72], and still another, when five months old, on hearing the rumbling of the cars in the street, knew to which window to go to look for them [11]. Schultze observed that active hearing, with attention, began after the

---

[1] Cf. [67] p. 109, where it is recorded that a child during her second month began to look at the piano keys as the source of the sound.

first half-year. Not only are there these differences among different children, but in the same child the accuracy of localization becomes greater by exercise. The differences in time, noted above, are doubtless in part due to variations in the rapidity of the physiological development of the ear.

By the end of the fourth month the normal child has made considerable progress in the *understanding of the meaning of sounds*, i.e., in the interpretation of sounds by their *timbre*. I find here also great differences in the results of the observations. Tiedemann's son took notice of gestures on the thirteenth day. Words would stop his tears or call them forth, according to the tone in which they were uttered. Another child, sixteen days old, would sometimes leave off crying when his mother spoke soothingly to him. At two months he distinguished between the loud bark of a dog and a coaxing yelp, being frightened by the former, but quickly soothed by the latter. A girl of three and a half months "knows when she is being scolded" [66]. On the other hand, out of one hundred children observed, Dr. Demme found only two who, at three and a half months, knew their parents' voices [72]. Another observer reports that at two months there was no apparent appreciation of ordinary sounds, but children of four and a half months sometimes recognized a voice [13].

These differences are, no doubt, to some extent, due to heredity, and to some extent produced artificially in the life of the individual by exercise. The average child apparently begins to comprehend the meaning of tones from the second to the fourth month.

A very interesting point in connection with the subject of the child's hearing, is his *power to appreciate music*. So intimately associated is it with the development of his æsthetic nature, that it deserves the careful study of the psychologist and the educator.

There are two chief sources of pleasure in music: the rhythmical movement, and the melody — the time and the tune. With regard to the first, it seems safe to say that no healthy, normal child, after the first few weeks, fails to appreciate rhythmical movements. At sixteen days one boy was soothed by the gentle, regular movements of the mother. These first musical impressions have a physiological explanation. There seems almost to be a *sense* of rhythm. The succession of notes produces a flow of blood to the brain, and its energetic excitation redounds in lively sentiments and animated movements. Thus music responds to that need of muscular activity so strong in the child. The social instinct also enters here: the child takes more delight in noise and movement when some one is at hand to participate.

With regard to the second point, the opinion may safely be ventured that no healthy, normal child is entirely lacking in musical "ear." I find no record of any child, who has been carefully observed, being utterly deficient in appreciation of musical harmonies. In the vast majority of cases the opposite is the case. Children almost always, from a very early age, show a lively interest in music. In one observed case, a child of one month manifested delight in singing and playing [66]. Sometimes children only two weeks old have been observed to stop the motions of their limbs, and apparently listen, when a piano was played in another room [100]. From six or seven weeks onward, and especially in the latter half of the first year, the child's pleasure in music is often shown by a sort of accompanying muscular movements, which he seems unable to repress. The mother's song of lullaby is keenly appreciated, and somewhat later is even given back by the child in a most charming infant warble. The emotional element in the music is often keenly distinguished. Dr. Brown says of one of the infants observed by her in New York city, that when only five and a half months old, he would cry when his mother played a plaintive air; but would stop at once, and begin to jump and toss his arms about and laugh, if she struck into a lively melody. There seems to be, as some one has said, a sympathy between the ear and the voice which antedates all experience, and which is even to a large extent independent of normal brain-endowment. Even idiotic children (provided they are not deaf) who can speak only a few simple words and syllables, are able to sing, and in singing they employ other words besides those generally at their command. While all this is true, it should also be remembered that the child's cerebral and mental endowment is exceedingly plastic, and that consequently sounds which at first were disagreeable to him soon become tolerable and even pleasant. He accommodates himself to all sorts of noises with far greater facility than the adult, and soon comes to take great delight in any sort of rude, banging, grating sounds, especially if they are his own production. Hence there is no sense in the education of which greater care should be taken than the sense of hearing. As already said, probably all normal children are born with a capacity for musical appreciation, though of course not all in the same degree. Now in the early period — during the first four or five years of life — it is very easy to cultivate this musical capacity or to destroy it. If the child hears, every day, rasping, grating and discordant noises, he will come very soon to like these as well as the most harmonious. It lies within the power of parents and teachers so to cultivate the child's capacity in this respect as to minister in an incalculable degree to the happiness of his life and the purity of his character.[1]

[1] "Comme l'a dit si bien le poëte, l'oreille est le chemin du cœur. Envelopper l'enfant d'une atmosphère de sons doux, tendres et réjouissants, c'est travailler à son bonheur actuel, et c'est faire beaucoup pour son humeur et sa moralité futures." (65).

III. TOUCH.

Touch has been called the universal sense, because, while sight, hearing, etc., have each a special, local end-organ, touch has its end-organs in every part of the body, numberless nerves of this sense communicating with the brain from every portion of the skin. The importance of the touch-sense is, therefore, obvious. Some have gone so far as to call it the fundamental sense, and have endeavored to reduce all the others to it. Without going this far, we may readily recognize its importance in the mental development of the child, from recorded cases of children who, from birth or from an early age, have been deprived of the other senses, or the most important of them, and who have, nevertheless, almost by touch alone, reached a remarkable degree of intellectual and moral attainment[48]. The field of the present inquiry is covered by three questions:

(1) As to the first beginnings of touch experiences (2) As to the comparative delicacy of different parts of the body. (3) As to the education of touch perception.

(1) All observers concur in the opinion that the sense of touch is exercised to a considerable degree in the fœtal stage of existence. Cabanis expressed the opinion that the sense of touch is the only one that furnishes the child in the first days with distinct perceptions, "probably because it is the only one that has had any exercise before birth." Kussmaul believes this sense is aroused in the embryonic period by contact with the surrounding matrix. Perez holds that there are indistinct tactile sensations during the intra-uterine life. Preyer believes touch-sensations are present at this time, though of far less intensity than in the subsequent life. Sully speaks of touch as the first sense to manifest itself. Erasmus Darwin expressed the belief that the fœtus receives through this sense some representation of its own

figure, and of the uterus itself. This opinion is concurred in by nearly all the authorities quoted in this connection here, and has been placed beyond doubt by the experiments of Kussmaul and Genzmer on prematurely born children, in whom they found the sense of touch already in full operation immediately after birth, though for a considerable time it is not accompanied by clear and definite objective reference, but is only a subjective feeling.

(2) Differences in sensibility to touch impressions among the different parts of the body are not so great at first as they afterwards become. In the uterus, the surrounding medium has been homogeneous; but from the time of birth onward, it becomes more and more varied, so that those parts of the body which are exposed to contact with the external world become relatively blunted in delicacy, while those which continue to be more or less protected — such as the eye and the tongue — retain more nearly their original sensitiveness. Nevertheless, the differences in delicacy among the different parts at the very first are surprisingly great.

The upper surface of the *tongue* is exceedingly sensitive. Kussmaul introduced a small glass rod into the mouths of children just born, eliciting prompt responsive movements, which varied in character according to the part touched. When the rod touched the tongue near the tip, the lips at once protruded, the sides of the tongue curled up around the rod, and sucking movements followed. When the rod came into contact with the back part of the tongue near the root, all the responsive movements — expression of face, mouth motions, etc. — indicated "nausea." (Similar results were obtained by Kroner and Genzmer.) No doubt we have here a sensori-motor reflex established before birth. The same is true in the case of the *lips*, which share with the tongue an extreme delicacy from the first. Even the lightest touch of a feather produced sucking movements of the lips on the sixth day [72], and gentle stroking of the lips produced the same result on the fifth day [47], and even on the first day [45].

One of the most sensitive parts of the body to touch impressions is the mucous membrane which lines the *nostrils*. This was observed to be sensitive on the first day of the child's life. "Tickling of the inner surfaces of the wings of the nose with a feather calls from children first of all winking of the eyelids, stronger and earlier on the tickled side than on the other; if the irritation be increased, the child not only knits the eyebrows, but moves the head and the hands, which latter it carries to the face" [47]. It appears, however, from the observations of the same authority, that this sensitiveness of the mucous membrane is formed only towards the end of the period of gestation, since similar experiments made on children born in the seventh month were without result.

Certainly next in order of delicacy — if indeed they should not have been placed earlier — come the various parts of the *eye:* the lashes, the conjunctiva and the cornea. Of these three, the lashes are considered by Kussmaul and Kroner the most sensitive to touch impressions. The former says: "The eyelashes are extraordinarily sensitive to even the faintest disturbances. If the child, when awake, has the eyes open, one can press with a glass rod even to the cornea before it will close the eyes; but should only one of the little lashes be disturbed in the least, this closing of the eyes will take place at once. The disturbance of the eyelids is not so efficacious by far; it will by no means be answered every time by eye-winking, as in the case of the cilia." He goes on to say that if one should blow through a small tube of twisted paper upon the face of an infant, winking will take place only when the stream of air has disturbed one of the cilia. Genzmer and Preyer differ from

Kussmaul here, holding that the cornea is more sensitive than the lashes. These facts are interesting as bearing on the question of priority between sight and touch in the eye. It has been frequently noticed that the child does not for a good while *blink* when a finger is thrust *at* the eye, provided it does not come into contact with it. Touch-reflexes seem, therefore, to be developed earlier than sight-reflexes.

If the tip of the *nose* be touched, both eyes will be shut tight. If one side be touched, the child will generally close the eye on that side. If the irritation be increased, both eyes will be closed and the head drawn somewhat back. This is an inborn defensive reflex.

If one tickles the *palm* of the hand of a new-born child, the fingers will close round the object with which it was tickled[47]. The skin of the face seems even more sensitive still. On tickling the sole of the foot, active reflex movements follow, such as bending the knees and hip-joints, curling and spreading the toes, etc. The reaction time is longer, however, in infants than in adults, sometimes amounting to two seconds. Slaps also are more effective than pricks, some children showing comparative indifference to the latter. A greater number of nerve ends are stimulated by a slap, hence the more speedy reaction. The greater sensitiveness of the adult to sense impressions in general is due to his more advanced cerebral development, and not to any superiority in cutaneous or nervous adjustment.

The other parts of the body are, speaking roughly, sensitive to touch impressions in the following order: The auditory canal (in the second quarter of the first year, the child observed by Preyer would instantly stop crying and become very quiet, if one's little finger were placed gently in the ear cavity), forearm, leg, shoulder, breast, abdomen, back, and upper part of thigh.

(3) The susceptibility of the sense of touch to education is very great, as may be seen from the attainments of those who are born blind, the proficiency they attain in reading by touch, etc. As a knowledge-giving sense, it stands very high, contributing much to the child's first knowledge of the external world, and, together with sight and the muscular feelings, to his first comprehension of space and time relations. It aids greatly also in his acquirement of the notion of self — this probably at first through touching some portion of his own body, and then some external thing, and feeling a difference between the resulting sensations [67]. But even before active touch has thus begun, the foundations of the child's education are laid in passive touch experiences, which from the beginning not only yield him pleasure and pain, but, being more frequent as well as more varied in their operations, contribute earlier and more largely than any of the other sense experiences to the development of his faculties, and to his gradual acquaintanceship with the world of objects by which he is surrounded.[1]

IV. Taste.

According to Sigismund, taste is the first of all the senses to yield clear perceptions, to which memory is attached. Not only is the exercise of this sense connected from the first with the child's most primitive needs and their satisfaction, but it is more than probable that, even in the embryonic stage, taste has been to some degree aroused by swallowing the amniotic fluid.

Numerous careful experiments show that the child is capable of *bona fide* sensations of taste in the earliest

---

[1] On this subject see Perez, "Education Morale dès le Berceau," Chap. V.

moments of life; and that, though he is for some time more obtuse and more uncertain in this respect than the adult, yet when a sapid object is introduced into his mouth, the resulting sensation really takes place by way of the gustatory bulbs and nerves, and is not merely a species of touch sensation, as some have held.

Kussmaul experimented on twenty children, during the first day of life — some of them in the very first moments — with the following results: Solutions of sugar and of quinine being introduced into the mouth by means of a hair pencil — the mixture being warmed so that the feeling of temperature should not affect the result — the children responded with "the same mimetic movements which we designate among grown people as the facial expressions of sweet and bitter." They responded to the sugar by protruding the lips in a spout-like form, pressing the tongue between them, sucking and swallowing. On the contrary, when the quinine was introduced, the visage was distorted, the eyes closed, the tongue protruded, and choking movements were made, accompanied by the expulsion of the fluid and active secretion of saliva. "Sometimes the head was actively shaken, as in the case of grown people when attacked by nausea." These results were obtained also in premature children, showing that this reflex arc is capable of performing its functions before birth. He adds, however, that he found great individual differences among children, some being far less responsive than others. Sometimes also the children seemed to make a mistake at first, as they occasionally responded to sugar by the mimetic movement for bitter, but this was probably only surprise at the *new* sensation, as they very soon changed it for the correct expression. He found also by these experiments that only the tip and edges of the tongue represent the tasting compass, the middle of the back part yielding no sensations of taste.

Genzmer, experimenting on twenty-five children, most of whom were just born, obtained results substantially agreeing with those of Kussmaul. He noticed, however, that in many cases the introduction of an attenuated solution of quinine was responded to by sucking movements, while stronger solutions were rejected with the mimetic for "bitter," showing that taste sensibility is weaker at this age than in the adult.[1]

Preyer agrees with the above deductions in every respect, and adds: "It is certain from all observations that the newly-born distinguish the sensations of taste that are decidedly different from one another,— the sweet, sour and bitter" (72). His boy, on the first day of life, licked powdered cane sugar, whereas he licked nothing else. Later, on receiving a strange food, he often shuddered and distorted his face merely on account of the novelty of the sensation, for, in the case of an agreeable sensation, he directly afterwards desired it, and received it with an expression of satisfaction. He concludes that the association of certain mimetic contractions of muscles with certain sensations of taste is inborn.

The development of taste-perception in the infant is interesting and important. The pleasures and pains of taste play a large part in his early education. The mouth is soon made the test organ to which all objects are carried, and by which their qualities are ascertained. Preyer's boy, on the second day, took without hesitation cow's milk diluted with water, which, on the fourth day, he stoutly refused. During his sixth month, he began to refuse to take the breast (which was offered him only in the night), because the sweetened cow's milk, which he had taken in

---

[1] These results are corroborated also by Kroner, Fehling and several others.

the daytime, was somewhat sweeter. From this time onward, and especially after weaning, his discrimination became much nicer, and by the fourth and fifth years he had become so "fastidious" that even the sight of certain articles of diet would call forth from him the mimetic movements for nausea, choking, etc.

Perez says the sense of taste is very slightly developed in the new-born, yet it exists. A child observed by him distinguished milk from sweetened water, and sweetened water from plain water, by the taste. Yet there are great differences of gustatory sensitiveness among children. In some cases, a child of six months has been induced to take bitter medicine by a change in the color. On the other hand, a child of two and a half months refused its bottle because the milk was not sweetened. Most children begin very early to detect the acid taste in certain substances.[1]

Yet in general, children's tastes change very easily, and hence are highly susceptible to education in almost every direction. Moreover, there are differences in the same child at different times: the state of the health, the temperature of the food (which, according to Champneys, is of more consequence than the taste itself), and many other circumstances entering in to disturb the gustatory equilibrium.

V. Smell.

Taste and smell are so closely associated that they might almost be considered together. The savour of substances depends, to a large extent, on their odor. These senses resemble each other in the comparative diffuseness of their perceptions, and in the fact that their sensations are more

---

[1] Dr. Brown thinks this is the first taste to be recognized.

persistent, and, therefore, less clearly distinguishable successively than those of the higher senses.

In order to sensations of smell, there must be air in the nasal cavities; hence there can be no exercise of this sense before respiration begins; none, therefore, before the beginning of the post-natal life.

Careful tests upon new-born children, however, show that they are susceptible to *strong* odors in the first hours of life. Records are at hand of tests made on about fifty children, most of whom were less than a day, some only fifteen minutes old. The tests were made with asafœtida, aqua fœtida, and oleum dipelli. Care was taken to experiment on sleeping as well as waking children, in order to avoid mistakes in interpreting the gestures and facial expressions. The result was that the children became uneasy, knit the eyelids more firmly together, contracted the muscles of the face, moved the head and arms, and, finally, awoke, sometimes even with crying. On the removal of the odor, they would fall asleep again. These results were also obtained in the case of eight months children, but not on those of a still more premature birth [47].

With the child's growth, progress is normally made in power of discrimination by the sense of smell, though more slowly than in the case of the higher senses. A little girl of eighteen hours obstinately refused a nipple on which a little petroleum had been rubbed, but readily took the other. Another child refused cow's milk *when it was brought near him*. Another, at thirteen days, refused certain medicines, being guided solely by their odor. Decisive discrimination of pleasant from unpleasant odors, with rejection of the latter, and appreciation of the former, has been observed in numerous instances from the early part of the second month on; and during the second half of the first year, this discrimination has become, with some children, very marked indeed, a lively enjoyment of the scent of flowers often being noticeable from this time on.

With all this, however, the sense of smell is far less acute in children than in adults. They often appear unaffected by odors which would be exceedingly unpleasant to the grown person. Further, their sensibility to smells very quickly becomes blunted by repetition or continuance, as is the case, to a less degree, with all persons. When the experiments with asafœtida, etc., described above, were repeated, no responses could be elicited after the first or second trial. Even after the child has become keenly appreciative of odors, he seems utterly to lack that dexterity in the management of the organ which is so noticeable in the case of taste. Children well on in the second year of life may be observed to carry a fragrant flower to the mouth — and even into it — instead of to the nose. The same awkwardness is seen in the management of the breath. When learning to smell, they invariably exhale with great vigor at first, but require considerable practice before they can inhale the odors.

Man seems greatly inferior to many of the lower animals in regard to smell. A kitten, three days old, "spat" at a hand which had been licked by a dog — a remarkable instance of the persistence and transmission of what Mr. Darwin calls "serviceable associated movements." The keenness of scent in dogs and horses, and many wild animals, is proverbial. In man, on the other hand, this sense stands very low in the knowledge-giving scale. Even in mature life, it gives but little information respecting the external world, and that of an uncertain character. In the child, it is concerned chiefly with the recognition of food. But it may well be that if this sense were brought into as constant requisition as the sense of sight or hearing, and as much care bestowed upon its education, very important results might take place in the way of developing a smell-sensibility.[1]

VI. Temperature.

There are two classes of thermic sensations: 1st, passive, subjective and general, as when we say "I am cold" or "I am warm." 2d, active, objective and local, as when we touch a hot or cold object and pronounce it hot or cold. Both are important in the child's development, but the second sort lends itself to experiment more readily than the first.

The sense of temperature should not be confounded with the sense of touch; for, though, like touch, it is universal, having its end organs scattered all over the body, yet the feeling in the one case is quite distinct from that in the other.

With regard to the possibility of sensations of temperature prior to birth, Luys expresses himself as follows: "We know indeed that from this period (the fourth month of pregnancy) the fœtus is sensitive to the action of cold, and that we can develop its spontaneous movements by applying a cold hand to the abdomen of the mother." Perez also is of the opinion that the fœtus experiences certain cutano-thermal sensations from about this time. Preyer takes the opposite ground, arguing for the homogeneity of the uterine temperature, and the consequent absence of any possibility of comparing sensations.

At all events, in the newly-born, the sense of warmth and cold develops very promptly. The gradual cooling, on coming into contact with the external world, the atmosphere,

---

[1] Mantegazza complains that we aid our eyes with spectacles, microscopes and telescopes, and our ears with trumpets, while the nose is entirely neglected. "Die Hygiene der Sinne."

the clothing, the bath,—all contribute to the speedy differentiation of thermic sensations, and to the perception of temperature. Genzmer, in experimenting upon about twenty new-born children, found that there was active withdrawal of the parts — palm of hand, sole of foot, cheek, etc. — to which the cold object was applied. His experiments are not entirely satisfactory, however, since sufficient care was not taken to exclude touch sensations from participating.

Satisfactory observations as to the development of the temperature sense are very scarce. Preyer found that the warm bath was enjoyed almost from the first, but the cold bath was disliked until the child learned by experience its refreshing effects. The lips, tongue and mucous membrane of the mouth were surprisingly sensitive to warmth and cold, even in the first days. The child would refuse milk of a temperature only slightly higher or lower than that of the mother. Still, on the whole, the infant suffers less from extremes of temperature than the adult, in whose case the faculty of judgment enters to aggravate the sensation.

An interesting point in this connection is the gradual variation between the "neutral point" in the tongue and cavity of the mouth, on the one hand, and the external parts, such as the hand, on the other. In the former it remains through life almost the same as before birth, while in the latter it gradually lowers by contact with the surrounding medium.

VII. Organic Sensations.

By this is usually meant those comparatively vague and general feelings of comfort and discomfort arising from certain conditions of the viscera, as distinguished from definitely located feelings resulting from excitation of the special sense organs. Hunger and thirst may serve as examples of visceral discomfort, and the feeling of satiety that follows the taking of nourishment as an example of visceral comfort. We shall also consider here feelings of pain in general, whether produced by external or internal stimuli.

The question of the possibility of pain experiences before birth may perhaps be considered settled by Preyer's investigations on fœtal guinea pigs and dogs (see "Physiology of the Embryo"). He obtained reactions which showed this sensibility to be present. The reactions, however, were very much slower than in the subsequent stages of life; showing either that the sensibility to pain is much lower in the fœtal stage than subsequently, or that pain reflexes are not firmly established at this time. Other investigators have found indeed that in the case of the very immature fœtus, the prick of a pin produced no response, although in the mature child, distinct reactions took place, by cries and movements, to strong mechanical or electrical stimulation.

The fact that the new-born child is capable of pleasure and pain also corroborates the view that his physiological apparatus is already adjusted before birth to this sort of experience.

Kussmaul has made some observations which go to show that very soon after birth, from the sixth hour on, but varying much in different children, the infant " is accustomed to betray distinctly that it is visited by a sensation which we must interpret as hunger or thirst, probably a mixture of both." This feeling is expressed by uneasy motions of the head and hands, sucking movements, and crying. One child, in the sixth hour of her life, would turn her head with surprising quickness, first to one side and then to the other, in order to take into the mouth and suck the finger with which the observer stroked her on each side of her face in succession, though he took care that in stroking the finger should not touch her lips [47].

Preyer observes that hunger and thirst assert themselves in sucking movements from the first. Very soon the cry of hunger is distinguishable from the cry of pain, being carried on with more intervals and in a lower tone, while the tongue is held in a peculiar manner, being drawn back and spread out. The hungry infant he also observed to move its head from side to side in a way not seen in any other circumstances. Gradually the child becomes relatively less absorbed in the satisfaction of hunger. From the fifth month, he can be diverted from eating by new noises and movements. From the tenth month, his eating is not so hurried and greedy. This is partly owing to the fact that at this age he takes more food at a time, the stomach being very much larger than at first.

For the rest, but few observations have been made. The child experiences organic sensations of pleasure and pain (the pain possibly predominating in the earliest period) in connection with the digestive, respiratory and circulatory processes: pleasure in their normal functioning, pain when the organs are fatigued or diseased. Pleasures in general are expressed by the widely open and "swimming" eyes, by the smile,— which, according to Darwin, occurred for the first time as a *real smile* on the forty-fifth day,— and by "crowing," joyful tones of voice; pains by tightly closed eyes, mouth drawn down at the corners, and later by the quadrangular form of the mouth in crying, while the cry itself varies according to the cause. The child is much more easily fatigued than the adult, and during the first few days passes most of the time in sleep.

VIII. Muscular Feelings.

We assume that in the normal condition all muscular movements are accompanied by muscular feelings. It is a sort of "internal touch" spread all over the body, and intimately associated with locomotion and prehension, with expansion and contraction, with pressure, weight, resistance, etc. It also includes the "feeling of the state of the muscles when at rest." So closely connected with the child's activity, its bearing on the rise of will is obvious.

That the child's muscles are called into play during the later months of his ante-natal life, in a great variety of movements, is so fully established as to require here only a passing word. It has been supposed by some that the fœtus is incited to muscular movements by the tedium of his unchanged position. It seems better, however, to suppose that now, as at a later time, there is an instinctive necessity for movement. The child is exceedingly active. To move his muscles is for him an absolute necessity, and the wisest methods in child training are those which recognize this fact, and, instead of repressing his activity, direct it into the best channels.

Though muscular feelings are present thus early, they are probably very vaguely apprehended by the child during the first month of his life. By the end of the third month, however, a vast number of these feelings have become associated with visual sensations, by means of coördinated movements of the neck, arms and eyes. About this time also begins the discernment of weight, though the appreciation and comparison of different weights are probably later attainments. The healthy child experiences the keenest pleasure in the exercise of his muscles. One observed case may stand for many. A little boy, in his fourth month, was observed to hold his toy rabbit up by the ears, crowing proudly, in evident enjoyment of the effort [11]. It is likely, as Ferrier says, that the muscular feeling of effort, by which weight is discerned, is first discriminated in connection with the movements of respiration.

From about the middle of the first year, the healthy child develops a remarkable propensity to seize, lift, pull, and otherwise handle all objects that come within his reach. This is to be attributed partly to natural curiosity, but more particularly at this early period to the constitutional need of exercising the muscles, to which he yields almost unconsciously. As soon as he is able to walk, the range of his muscle-activity is vastly extended, and from this time forth, his experiences in this connection play a large and important part in his education.[1]

[1] For further remarks on muscular movement, *vide infra*, Chap. IV.

CHAPTER II.

EMOTION.

The principle of transformation, which is exemplified in almost every fact recorded in the preceding chapter, is still more clearly illustrated in those departments of the mental life which we have yet to consider. In studying the emotions of children, for example, we shall observe that in the earlier stages, when intellectual comprehension (which is essential to the emotions of the grown-up person) can by no means be presumed to be present, yet the outward manifestation — movement, facial expression, etc. — resembles very closely that of the adult, or the older child. It seems unphilosophical to class the phenomena of these two periods together under a common name, and our only excuse for doing so is that the one shades off so gradually into the other that to establish a rigid line of distinction seems impossible. We shall, therefore, consider both the stages under the head of emotion, only premising that, in the absence of active thought, these appearances can only be accounted for as the response of the organism to pleasurable or painful feeling. But later, when the mind asserts itself, and the human being begins to understand the cause of the feeling, and to interpret the gestures of others as the expression of *their* feelings, emotion, in the strict sense of that word, arises. The same physiological expressions continue to be employed, because through habit they have become easier than any others, while their employment in the first stage may be accounted for on the principle of heredity.

I. Fear.

These remarks are specially true in the case of fear, whose manifestation is at first quite independent of thought, and of specific experiences (as in the case cited by Perez of convulsive tremblings, even in the fœtus in certain circumstances), but which, as a true mental phenomenon, requires both these for its full development.

We have, then, two stages of fear: First, the fear that is independent of hurtful experiences, and must be considered hereditary; and secondly, the fear that is produced by a mental image of the danger. The former is very marked in the lower animals. When Spalding let loose a hawk suddenly over a brood of young chickens in a meadow, they immediately "crouched" and hid themselves in the grass, while the mother hen attacked the foe with tremendous violence, though neither she nor her brood had ever seen a hawk before. A dove, let loose in the same way, produced no such result. So the child, when only a few weeks old, will start and cry at any sudden sound or strange sight, quite independently of experience. He shrinks from cats and dogs, without ever having been injured by them; he is afraid of falling, before he has ever fallen, and trembles at the sight of large and majestic objects, such as the ocean, when he looks upon them for the first time [58]. Many infants cry when it thunders, though they do not at all understand what it is, and experience a shock — just as some nervous adults do — when a door closes with a bang, or an object falls upon the floor. They contract all the muscles of the body nervously when suddenly lowered through the air in the nurse's arms. They sometimes shrink from people dressed in black, and from those who speak in deep sepulchral tones. A little girl, slightly over two months old, appeared terrified on beholding a distorted face; she cried out, and sought protection in her mother's arms. "It was long before she was restored to her accustomed tranquillity — the vision reappeared in memory, haunted her fancy, and brought tears to her eyes"[100]. A child of seven months seemed afraid when a fan was opened and closed before him; another at a loud snoring noise which he heard for the first time. A boy of ten months was frightened by a squeaking toy; he soon, however, became accustomed to the sound, and even took pleasure in making it squeak himself[15].

In this early period, most children seem more afraid of *sounds* than of *sights*. Sigismund says fear develops from the time of the development of the ear. They are usually afraid of thunder, but scarcely ever of lightning. A child who started nervously when a box of comfits was shaken before him, made no such sign when the empty box was shaken[21]. One may thrust with the finger, as we have seen, quite close to the open eye of an infant, without causing him to blink, while, if one speaks to him in a harsh or loud tone, he will cry. A little child has been known to lie smiling in his cradle, surrounded by the flames of a burning house; but when rescued, has broken out into loud cries of fear at the noise of the engines and the shouting of the assembled crowd.

Eye-fear, however, soon develops, and strange sights as well as sounds startle and frighten the child. We have a very ancient example of this in the Iliad, where Hector is described as bidding his wife and child farewell before going out to the fight. When he reached out his arms for the child, the latter cried out, and hid his face in the bosom of the nurse, frightened by his father's gleaming bronze, and the helmet crested with horse-hair. Sigismund describes his child as showing fear of a sleeve board, by association with the glowing "goose," and also at the sparks from a blacksmith's forge. There are also touch-fears. The little girl F. started back when her hand came into contact with some soft fur. The suddenness of the sensation apparently had more to do with her fear than the quality of the feeling, for she soon lost her fear of this article.

Quite different from all this is the fear shown by a child in the presence of an object which has, on some former occasion, caused him painful feeling. Preyer's boy, at nineteen months, screamed *at the sight of* the cold bath and sponge, from which he had, on a previous day, received unpleasant sensations. Here the *idea* causes the fear, memory coöperates, and child has become susceptible to fear in the strict sense. This probably might have been observed earlier.

The plasticity of the child's nature renders him susceptible to impressions which, in many cases, remain with him through life. Fear of the dark, fear of the woods, fear of being alone, are often inculcated by unwise nurses and teachers, and remain, in some cases, ineradicably fixed in the constitution. Mosso tells of an old soldier who, on being asked what had been his greatest fear, replied: "I am nearly seventy years of age. I have looked death in the face many times, and never felt fear; but whenever I pass a little church in the shadow of a wood, or a deserted chapel in the mountains, I always remember an abandoned oratory in my native village, and am afraid. I look around, as if I were about to see the corpse of a murdered man which I saw in my infancy, and with which an old servant threatened to shut me up in order to quiet me."

The child from three to seven years is very liable to have dreams of exceeding vividness, and if he wake suddenly out of a deep sleep, his face will often bear signs of great fear, as though he saw an apparition. The eyes stare straight ahead, he fails to recognize persons, he breaks out into perspiration, his heart beats hard and his limbs tremble. These nocturnal fears may become so strong as to cause veritable attacks of epilepsy [58].

Sometimes a new fear is developed by sickness. Some children seem morbidly timid and fearful, while others seldom show signs of fear in any form. As the child's education progresses, his fear increases in some directions, and decreases in others; as he learns, on the one hand, that certain objects which he supposed harmless are really harmful, and on the other, that some which he at first esteemed dangerous, will do him no injury. In other words, it is only a commonplace to say that fear is both increased and diminished by advancing knowledge. The man is *more* afraid of a *loaded* pistol, and *less* afraid of an *empty* one, than the child.

II. Anger.

Anger (which, according to Plato, is one of the natural attributes of the soul, and closely akin to courage) is evil only in its abuse. In a moderate degree, it is the index of a just and sensitive temperament, and a force which education should *direct* and not *annihilate*. "In my opinion," says Perez, "a child of ten months who does not weep or cry at least four or five times a day, who is not amused, and who is not irritated, like a savage, or a young animal, by a mere trifle ("pour une bagatelle"), is lacking in sensibility and in intelligence, and will, no doubt, be lacking in character,—bury him; he is dead." "It is necessary," he goes on to say, speaking of the education of the child in this regard, "to surround the cradle with an atmosphere of sweet serenity, but it is not always necessary to hide anger. Just anger should be shown, but with moderation" [65].

It is difficult to say when the child first feels anger,

because its outward signs are at first very easily confounded with those of pain or distress. Mr. Sully thought he saw manifestations of anger at the very outset of life, in a little girl, who, " in refusing to accept the nutriment provided by nature, showed all the signs of passionate wrath." Mr. Darwin noticed, in a child eight days old, frowning and wrinkling of the skin around the eyes before crying; but he adds, "this may have been pain and not anger." In the third month, he thought he observed signs of real anger, and in the fourth month he had no doubt about it, for the blood rushed into the face and scalp. Tiedemann's son gave evidence of anger in the second month by actively pushing away the disagreeable object. By the eighth month, he was quite capable of violent anger and jealousy. Perez believes he has seen signs of impatience at the end of the first month, if not earlier; and, in the second month, real fits of passion, pushing away distasteful objects, frowning, reddening, trembling and weeping. At six months, children will scream if their toys are taken away, and towards the end of the first year, anger sometimes exhibits itself in revengeful actions hurtful to themselves, such as beating a chair, etc.[13]. A child of seven months screamed with rage because a lemon slipped out of his hand; and at eleven months, if a wrong plaything were given him, he would push it away and beat it [21].

Up to a certain age, almost all children are exceedingly irascible, and I know of no particular in which the familiar analogy of the child to the savage is more strikingly shown. The child is a little savage. His will and reason are weak, his passions are strong, comparatively speaking, and he is ruled by his feelings. So it is with savage races. They are proverbially passionate; and the progressive effects of civilization upon a race, leading them gradually to control the impetuous and unreasonable rage which characterized the earlier stages of their civilization, is strikingly analogous to the wise training of the human being from the irascibility of the child to the calmness and moderation of the educated man.

III. Surprise, Astonishment, Curiosity.

Surprise and astonishment are closely related to fear; novelty of impression and failure to understand being the underlying causes in all three.[1]

Surprise and astonishment are not identical. The former may be described as an active state, the latter as a passive one. The child who is only surprised maintains control of his muscles, and examines the strange object with the closest attention, while the astonished child suddenly loses volitional control, and remains fixed in the attitude in which the strange impression overtook him, with wide-open mouth and eyes. In the one case there is activity and movement, in the other a sort of paralysis.

Surprise has been observed in a child one week old, who stared at his own fingers with great attention. Doubtless he had never noticed them before [72]. From this time onward, wonder is constantly manifested at pictures on the wall, sunbeams dancing on the floor, the fire crackling on the hearth, and especially at the movements of animate beings. The infant gazes long and steadily at these strange phenomena. A little girl of less than a month, on being taken downstairs into new quarters, stared round in great wonder for a time, but this soon passed away [100].

---

[1] "The most powerful agent in the development of the understanding at the beginning is astonishment, together with the fear that is akin to it." Preyer. "Sometimes wonder passes into awe, or even fear." Sully.

Astonishment makes its appearance later. The following are Preyer's observations on this point: In the twenty-second week, the child was struck with astonishment when his father suddenly appeared and spoke to him while they were riding in a railway carriage. In his sixth and seventh months, the same thing occurred at the sight of a stranger in the room. The child's eyes opened wide, his lower jaw dropped, and his body became motionless. In the eighth and ninth months, these symptoms were still more pronounced, but it was noticed that astonishment was manifested generally at sights and sounds, and not at impressions of taste and smell. The child manifested astonishment at the opening and shutting of a fan (31st week); at the imitation of the voices of animals (34th week); at a strange face (44th week); at a new sound (52d week), and at a lighted lantern seen on awaking (58th week). Along with the gestures described there was sometimes the sound of "ah," made by involuntary expiration of breath. By the end of the second year, these signs of astonishment became more rare, as the child grew more accustomed to strange sense-impressions.

It is to be observed that the peculiar manner of expressing this emotion, as well as most of the others, is entirely original with the child himself. He expresses astonishment in this way before he has had any opportunity of imitating the gestures of others. These gestures, therefore, must be the result of instinctive tendencies, which, by virtue of heredity, have become fixed in the human race, as they are everywhere the same [22].

M. Egger emphasizes the close relationship between the feeling of wonder and the religious sentiment, and holds that the child is by nature predisposed to religious ideas, whose germs he, in fact, brings into the world with him. M. Perez, on the other hand, following Spencer, maintains that there is no innate predisposition in the child to look beyond the natural to the supernatural, and that, apart from training and example, the religious ideas would never take root in his mind. In the absence of conclusive evidence on the point, all opinions must be merely hypothetical. It may, however, be suggested that if the familiar analogy between the infancy of the individual and that of the race is to hold here, we must accept M. Egger's position, since almost all savage races are deeply religious, abounding in ideas of the supernatural.

Closely allied to the sentiment of wonder is that of *curiosity*. This is a natural, spontaneous tendency, which might perhaps be more fittingly classed under the head of intellect, but for the fact that, in the very young child, its essential character is feeling. It consists of a sort of chronic hunger for new sensations, which impels the child constantly to handle, examine, taste, and otherwise experiment upon all objects that come within his reach. The little boy R. used to try to untie every parcel that was brought in. It is a purely sensuous impulse at first, but with the expansion of the intellect, it is transformed into the pure desire to know. It permeates the *play* of the child, which, as Sigismund says, is like the experimentation of the scientist, by which he elicits from nature the answers to his questions. It is one of the most powerful factors in the child's development, and should be guided into right channels, rather than discouraged, by the educator.

Tiedemann believed curiosity was developed in his son in his second month; the eyes made an effort to follow a new or curious object. Perez saw evidences of curiosity almost from the beginning, and at two months the child " would stretch out his hand, and turn his eyes and ears towards objects affecting his senses. At three months he would seize objects within reach, and shake them about to amuse himself." From this time on, and especially from the time he begins to walk, everything within reach becomes the object of constant study. The acquisition of language adds greatly to his resources in this respect. "His little voice, a hundred times in an hour, expresses a desire, or asks a question, and that, not so much through need of knowing what *things are*, . . . as through the appetite for fresh and new sensations. So powerful does this impulse become that sometimes the child is sad, or even sick, if it be not gratified" [65].

M. Taine calls attention to the significant circumstance that this curiosity, which is so powerful a force in child life, is not found in the lower animals. "Any one may observe that from the fifth or sixth month, children employ their whole time for two years or more in making physical experiments. No animal, not even the cat or dog, makes this constant study of all bodies within its reach. All day long the child of whom I speak — twelve months old — touches, feels, turns about, lets drop, tastes, and experiments upon, everything she gets hold of, whatever it may be — ball, doll, coral or plaything. When once it is sufficiently known, she throws it aside; it is no longer new; she has nothing further to learn from it, and so has no further interest in it" [98]. It will be noticed here that Taine assigns a larger part to the intellectual than does Perez. He says physical need and greediness count for nothing. It is pure curiosity. "It seems as if, in her little brain, every group of perceptions was tending to complete itself, as in that of a child who makes use of language." But the little girl observed by Taine was a year old, and by that time, no doubt, curiosity was beginning to assume more of an intellectual character.

IV. Æsthetic Feelings.

As early as the forty-fifth day, Mr. Darwin noticed a real smile of pleasure, "which must have had a mental origin." It was observed when the infant was looking at his mother, and also during the act of nursing; and was quite different from the so-called smiles which had been seen prior to that time, in being accompanied by a more intelligent expression, and by the sparkling and "swimming" of the eyes.

It is not to be presumed that every laugh of the young child proceeds from a comprehension of the *humorous*. The first laugh is probably — like the first vocal utterances — only the spontaneous functioning of the organism. Yet it is maintained by careful observers that the sense of fun is present in some children three months old [21]. About this age they may be greatly amused by such little games as throwing a pinafore over the head and suddenly withdrawing it, and by the familiar gambols of hide-and-peek. Later they show great pleasure at being carried on one's shoulder, swung about in the air, or tossed up to the ceiling. They laugh most heartily while the fun lasts, and are very unwilling that it should stop [C].

Something has already been said on the subject of *musical appreciation* in children. Mr. Darwin, who observed in his child a fondness for the piano as early as the fourth month, considers the feeling of pleasure in music as the first of the æsthetic sentiments, unless the appreciation of bright colors comes earlier. Another child, at five months, showed signs of pleasure when singing was going on, and even kept a sort of time with his body, but was indifferent to whistling [101]. Another observer places the pleasure in musical sounds as early as the second month, and in another case the child was observed at eleven weeks to pucker up his lip a little when the piano was being played [C]. I have frequently observed this fondness for music at a later age, when the child will crowd close to the piano, and show his appreciation by rocking his body to and fro. Appreciation of expression in music is, however, almost entirely lacking at this time, and requires education to develop it.

SENSE OF MATERIAL BEAUTY. — The child at first confuses *the beautiful* with *what is pleasant.* Animated movement at the sight of beautiful things is at first, no doubt, only response to pleasant feeling. There is no understanding of form, color, etc., as beautiful or otherwise. This pleasure, in certain sensations, however, is one of the foundation stones upon which the æsthetic sense of material beauty is afterwards to be built. From about the eighth month, there have been observed the beginnings of this feeling in the pleasure shown by the child in personal adornment. But even now the æsthetic and the sensuous are blended in the pleasure a child feels in the new dress or hat. "Pretty" and "good" are interchangeable terms in his mind. At thirteen months he will snatch at haphazard among a heap of toys, seeming not to discriminate at all among them as to beauty; and, at a much later period, a child taken out to the country gives no evidence of any appreciation of the beauties of the landscape, but is attracted rather by some new or strange object — especially if it be an animal, or something that moves. Symmetry in form and harmony in colors make but little impression on him. Here, as in music, he demands quantity rather than quality, movement rather than expression. Yet these words must not be understood as denying to the young child all æsthetic feeling. Beautiful objects, if they are not too large, nor too distant, please him. He is charmed by the pretty butterfly and the pretty flower; he is greatly attracted by the human face, and by the expression of the human eye.

The *dramatic instinct* is very strong in childhood, though stronger and earlier in some children than in others. Children are born actors. Their lively imagination and strong hereditary tendency to imitation lead them, even before the first year of their life has gone, to perform many curious movements and gestures. In their plays, children constantly personify, represent, dramatize, assume characters, and assign fictitious characters to other persons and things [63]. An eminent teacher in Toronto assures me that his three children, in their play, almost always address each other by assumed names, and the play is carried on in make-believe characters, which are dropped as soon as the game is over, and never referred to at any other time.[1]

V. Love, Sympathy, Jealousy, etc.

If we may judge by the smiles which an infant bestows upon those who have charge of him, *affection* for persons arises very early. These smiles have been observed before the end of the second month, and even at a much earlier period. The earliest smiles are probably automatic, as already said, but by the end of the fourth month there is no longer any doubt that persons are recognized. A little boy of this age was observed to lift up both arms towards his parents, "with an indescribable expression of longing" [74]. A girl of the same age used to be fond of lying beside her sister, their faces touching. After her sister died (she was then five months old), she seemed very lonely, and when she met other children of her own age, she would greet them with smiles and kisses [W]. In another case visible signs of

---

[1] It seems best to postpone any further remarks on this subject, until the imagination is taken up in regular order. See *infra*, Chap. III. sec. IV.

affection for persons whom he knew, were shown by a boy eight months old, and another boy, who, when nine months old, used to return his father's caresses by a charming smile and gentle stroking of his father's face, had grown very affectionate and sympathetic by the time he was fourteen months old, and bestowed his caresses in abundance, not only on his parents and friends, but on the cat and dog also [C]. Spontaneous expression of affection is, in many cases, indeed, first shown about the beginning of the second year. One child of this age kissed his nurse repeatedly on her return from a short absence, and another was in the habit of showing his affection for certain persons by gently laying his hand upon their faces or shoulders. Affection for animals, and even for inanimate objects, is also very strong in many children of this age. The little boy R. was remarkably attached to an old scarf of soft wool, and to a couple of rag dolls. He would not go to sleep without them, but would lie in his cradle and call for them until they were brought, when he would hug them up in his arms, and fall asleep chattering and cooing to them in a charming manner. When he got into any trouble, especially if his mother punished him, he would run and bury his face in the old scarf, and weep out his childish sorrows into its sympathetic folds.

The memory of the little child is comparatively weak, and his experience short; and hence, though capable of strong affection, that affection does not persist long in the absence of its object. "Out of sight, out of mind," is true in the case of the child during his first year, and relatively true to a much later period. He is incapable of "homesickness," with all its suffering, simply because he is unable as yet to form mental pictures of home and friends who are absent. He lives in the present rather than the past, in the realm of sense rather than that of memory. For the same reason, his love for persons and places is very plastic, and may be moulded and directed into almost any desired channel during these early months and years; hence the responsibility resting on those who are entrusted with his earliest education in home and school.

SYMPATHY. — There are two reasons why *sympathy* as a characteristic of childhood should be, during the first few months, so weak as to be almost entirely lacking. The first is that the child's life at this time is so full of his own personal needs that he can pay but little attention to those of others; the second, that he is as yet unable to comprehend the outward signs of feeling in others, because of the shortness of his own experience. It seems probable that some of the earliest manifestations of apparently sympathetic feeling may be merely the result of sensori-motor suggestion [7]. Sigismund noticed the first signs of sympathy at the end of the first three months, but Tiedemann says his boy, when only two months old, made sympathetic responses when consoled by the usual vocal expressions. Mr. Sully has observed the same thing. In another case a boy six months old drew a melancholy face, with mouth depressed, when his nurse pretended to cry [21]. At seven months, another child manifested decided altruism, and seemed desirous of sharing his pleasures — with the exception of food — with others. In another case a child of eight months cried when some one pretended to whip his nurse, and another child of nearly the same age made a mournful whining noise, accompanied by the facial expression of "crying," on hearing another child cry, and also when a minor chord was struck on the piano [B]. During the second year, sympathy becomes so strongly established that its outward evidences are sometimes seen, even on occasion of the imaginary sufferings of inanimate objects, and pictorial representation of suffering. A child of this age cried when her dolly was "hurt." Sympathy with human beings is, however, usually much stronger than animal sympathy. A child of one year, who returned home after a short absence, took no notice whatever of the cat or dog, but at once recognized his nurse and the other members of the family with pleasure. The strength of human sympathy, and the need of it in the child, is seen in the fact that when he is hurt, he rarely cries, unless there is some one near at hand to hear him.

JEALOUSY. — Children are naturally selfish and egotistic. It has been said that the "meum and tuum" are very much confused in the young child's mind. Perhaps it may be better said that his idea of "tuum" scarcely exists, while his notion of "meum" is enormously exaggerated. The proprietary instinct is very strong in some children, and this enters largely into the feeling of jealousy. "The need of play engenders the desire of possession" — i.e., of the playthings — and this in turn gives rise to the instinct of property; hence jealousy. Tiedemann's son did not want his sister to sit in his chair or put on his clothes, but he would freely take hers. "Jealousy depends in general on temperament, and is often the index of a very keen sensibility, though showing itself also in children of a calm disposition. It is easily confounded with envy, desire, wish to possess, need of being noticed, etc. It opens the way for hatred, falsehood, dissimulation; in certain feeble natures it leads to discouragement" [65].

"The child of three months shows by various signs a proprietary interest in the breast; handles it as his own, and is jealous if it be given to another. Later he demands it with still more 'authority'" [65]. "At three and a half months, little Mary is jealous in the extreme, and cries if her sister sits upon the mother's lap" [66]. From the eighth month another child gave every evidence of jealousy in similar circumstances; grew very angry, and tried to drive the usurper away. A little girl of ten months would cry "in a distressful way, not expressing anger, but disappointed desire, if the nurse took another child upon her knee." She would not be appeased except by being taken up. It would not do to take her on one knee, and the other child on the other; she must have sole possession (93). Little R. insists on being a sharer in any caresses that may be going forward between his parents. Darwin saw plenty of evidence of jealousy from the fifteenth month, and observes that it would probably be found earlier. So also Perez.

The jealousies of children need careful treatment. They are often augmented and rendered morbid by injudicious conduct, and thoughtless words of praise and blame on the part of grown-up people. Carefully treated, this feeling may be developed into self-respect on the one hand, and a proper altruism, or "jealousy for others," on the other, and thus contribute much to the child's moral education.[1]

---

[1] "In der Kindheit und am frohen Morgen des Lebens lebt der Mensch eigentlich nur sich selbst; da bildet sich durch 'Leben für sich,' der Körper und die Seele zum 'Leben für sich und für andere.'" (26)

CHAPTER III.

INTELLECT.

Most of the phenomena described in the preceding pages involve thought in a greater or less degree; yet in the earliest experiences, mental activity is at a minimum; the affective predominates over the presentative, and the representative occupies but a very small place. Yet it seems incorrect to say, with Nasse, that "mind comes first at birth, and the first breath is the earliest mark of intellect;" or with Heyfelder, that the first cry is the sign of awakening mind; or with Karl Vogt, that the newly-born possesses no trace of intelligence. Kussmaul seems nearer the truth in the following: "It cannot be doubted that man comes into the world with an idea — a dark one to be sure — of an outer something, with a certain idea of space, with the possibility of localizing certain touch sensations, and with a certain mastery over his movements. How can it otherwise be explained that the hungry child, before it is suckled, not only seeks nourishment, but seeks it in that region where its sense of touch during the search is actively excited? These astonishing actions can only be comprehended under the following suppositions: First, that the child has already gained the dim idea of an outer something which is able to remove the unpleasant sensation of hunger or thirst, and which, to that end, must come through the mouth; secondly, that he is able to decide the place from which the sensation of stroking came; and thirdly, that he has already learned to turn the head voluntarily to the one side or to the other." (47).

It is not possible, within the present limits, either to give a detailed exposition of the nature of the thought process, or to trace the intellectual development on into the maturer years. For these the reader is referred to the numerous standard works on psychology in general. Here we can only attempt to collate facts calculated to throw light on the first budding of the intelligence, and to trace each phenomenon only to that stage at which it may be said to be fairly "under way." The intimate relation between thought and language also makes it advisable to postpone much that might be said here, until we come to the consideration of the latter topic.[1]

Observation of intellectual development is hampered by two difficulties, which render great caution necessary. In the first place, the combined influence of heredity and environment produces such wide individual differences among children, that no general conclusions can be safely expressed until a very large number of cases have been observed. (Certainly nothing exhaustive or final can be said at the present time.) In the second place, even the most careful observer, watching *one child*, is apt to be misled by certain deceptive appearances, and to give the child credit for a good deal that he does not really know. "They do clever things, and say brilliant words, by imitation and accident, not knowing the meaning of them" (25). In this

---

[1] The relation of thought and language has perhaps never been more aptly expressed than by Sir W. Hamilton in the following: "Language is to the mind precisely what the arch is to the tunnel. The power of thinking and the power of excavation are not dependent on the word in the one case, nor on the mason work in the other; but without these subsidiaries neither process could be carried on beyond its rudimentary commencement." Lectures, Vol. 8, p. 138.

way many a child, supposed to be a prodigy, does not at all excel others, except in a quickness of imitation. When you want him to "show off," he fails you, simply because the words do not mean the same to him as they do to you, and his use of them is largely mechanical.[1] The child's act may resemble ours outwardly, but the sentiment underneath the act may be very different. G. S. Hall says: "Not only are children prone to imitate others in their answers, without stopping to think and give an independent answer of their own, but they often love to seem wise, and, to make themselves interesting, state what seems to interest us without reference to truth, divining the lines of our interest with a subtlety we do not suspect." In interpreting the phenomena here recorded, great care is necessary to avoid an inaccurate estimate of their intellectual value.

I. Perception.

In the process of perception — which may be simply defined as "that act of the mind by which real external things become known through the senses"[Y] — there are three stages, distinguished from each other qualitatively, though not chronologically. First, the simple feelings of the senses are differentiated. Changes, quantitative and qualitative, are felt and known. The child recognizes the difference between a sweet *taste* and a bitter one, for example. He could not describe the difference even if he could speak, but is simply *aware* of it. Secondly, the sensations are localized. A definite "whereness" is attributed to them. This involves the recognition of space properties

---

[1] As Rousseau says in Émile: "Un instant vous diriez: C'est un génie, et l'instant d'après: C'est un sot. Vous vous tromperiez toujours: *C'est un enfant.*"

in objects, and opens up the vexed question of the origin of the idea of space, into which we cannot enter here. Thirdly, the manifold of sensation, thus differentiated and localized, is unified into a permanent whole, which we call the *object*. The child combines the scattered sensations, visual, tactual, olfactory, and sapid, into the perceived object, *food*.

TASTE PERCEPTIONS. — "The first centre of the child's psychic life is the mouth " [36]. Probably the first action is sucking, and later all objects are experimented upon by means of the lips and hands together. But even in the third month, the child is weak in power of comparison, and will suck an empty bottle as readily as a full one, until he finds it is empty by failure to extract anything from it. From the eighth day, a wry face was made at the sight of bitter medicine, and by the seventh week this wry face was accompanied by a gesture of refusal [66]. At one month and five days, a dose of medicine was taken with visible repugnance [101]. The experiments of Kussmaul, already referred to, show that discrimination between tastes takes place from the first. It proceeds, generally, with considerable rapidity from the third month on, and by the tenth month various articles of diet are clearly known and distinguished from one another [66]. Yet the child, like the adult, though in a greater degree, is subject to illusions of taste, through confusion of sapid with olfactory sensations, and with one another.

SIGHT PERCEPTIONS. — During the first month, the child gives small evidence that he has any ideas of distance, or of his own body. At this age he will strike or scratch his own face. A girl of thirty days "seemed for an instant to have caught the reflected image of herself," but the next moment she became lost again in the surrounding objects of the nursery [100]. A boy, during his second month, gave the first sign of distinguishing external objects from himself, by reaching forward and grasping at them. About the same time he began apparently to pay attention to the looks and gestures of others, and at six months he distinguished persons, without, however, having any clear ideas about them. When anything presented itself to him, he pointed his finger at it, to direct attention to it, and sometimes said *ah* [101]. From the beginning of his second year, he rapidly advanced in power of discrimination, though chiefly among objects fitted to satisfy his needs. One of the objects earliest to be recognized — if not the very earliest — is the mother's face and form. Children give evidence of this recognition in the second or third month. A boy of seven months "surely recognized three persons," — his parents and the nurse [88]. Another, at nine weeks, seemed to know his mother [11]. No objects, not even the parents, are known at a distance [36]. In the course of the first half-year, much improvement takes place in this direction. A child in his fifth month would no longer grasp at objects beyond his reach [34]. Smiling at the image in the mirror has been noticed as early as the ninth week.

"From the sensations of *hearing* and *smell*, there can be formed no representations in the first week" [47]. Near the end of the second month, one child gave evidence that he distinguished between tones of voice expressive of different emotions and sentiments. He allowed himself to be pacified by gentle tones [101]. Another, in his third month, actively sought the direction of sound by turning his head [72].

Owing to the weakness of the attention, and lack of experience, the young child falls into many illusions of sense-perception. A child of four months believes the image in the mirror is a real person, as is shown by his surprised look when he hears behind him the voice of the individual to whom the reflection belongs [21]. A boy of seven months put out both hands to pick up a very small piece of paper [66]. At six months he mistook a flat dish for a globe, and seemed to believe all objects had bulk. The little girl F. tried one day to "pick up" a round picture, which was made to represent raised work, and another day she tried to walk on the water. I once heard a little girl of one year and a half call the moon a lamp, showing how false was her idea of its real distance and magnitude.

Children are said to be peculiarly subject to illusions of hearing, though I have no examples to give. The imperfection of their judgments by the muscular sense is shown by the fact that a child of three months cannot tell a full bottle from an empty one, by the weight alone.

II. Memory.

The power of retaining impressions, and recognizing them when reproduced, has a physiological as well as a psychological aspect; the former consisting chiefly in the susceptibility of organic structures to receive impressions which are capable of a greater or less degree of permanency; the latter depending principally on the power of attention. Where the attention is actively directed towards the present sensation, that sensation is more easily and more surely reproduced in memory.

Little children have but small power of attention; from the psychological side therefore, their memories are weak. Nearly all the experiences of the first two years of life, and the vast majority of those of the next four, are completely forgotten by most people.[1] The cerebral structures in chil-

---

[1] "A writer in a recent English magazine declares that her own memory began at sixteen months." M. W. Wright in *Babyhood*, Feb. 1891.

dren, however, are very impressible, so that, from the physiological point of view, the memory of childhood is potentially, at least, very strong. This probably accounts for the well-known fact that those experiences of childhood that *are* remembered, are more firmly fixed and persist longer than those of early manhood or middle age. Let the attention of a little child — which, be it observed, is weak in *both directions*, being as hard to *withdraw from* a present sensation as it is to *direct towards* one — be enchained by some startling or fascinating experience, and an impression is made on his plastic mind, which can never be effaced.[1] Old men recall the events of fifty years ago better than those of last year.

The little child is capable of memories long before he has learned to speak. A little boy, six months old, whose hand had been slightly burnt by a hot vase, shrank back at the sight of this article a few days after [25]. Certain faces, too, are recognized by children of this age, showing that they have memory-images of them. Strange faces, too, are known as strange, and distinguished from familiar ones; but the latter are not yet missed when absent [72]. Sigismund gives an interesting case of memory in a boy about eight months old. While in the bath he tried repeatedly to raise himself up by the edge of the tub, but in vain. Finally he succeeded by grasping a handle, near which he accidentally fell. Next time he was put into the bath, he reached out immediately for the aforesaid handle and raised himself up in triumph. Memory of persons becomes strong by the end of the first year. A child of this age recognized her nurse, after six days' absence, "with sobs of joy." A boy some

---

[1] My first sight of a locomotive will never, I believe, be effaced, or even bedimmed, in my memory, should I live for a century. To-day I can call it up with remarkable vividness, and with all its attendant circumstances clearly and definitely portrayed.

what younger knew his father after four days' absence. while another, seven months old, did *not* recognize his nurse after four weeks' absence, but when nineteen months old he knew his father, even at a distance, after two weeks' separation. Another child, four months old, knew his nurse after four weeks, and at ten months he missed his parents, and was troubled by their absence. A boy of twenty-three months manifested keen delight on again seeing his playthings after an interval of eleven weeks; and when a year and a half old, was greatly disconcerted one day when sent to carry *one* towel to his mother, where he had been accustomed to carrying *two* (C). Darwin's boy, at a little over three years of age, instantly recognized a portrait of his grandfather, "and mentioned a whole string of incidents which occurred at their last meeting, nearly six months previous," the matter not having been mentioned in the meantime. The little boy, R., recognized a young lady who lives next door, after a few weeks of absence. He also knew me after nearly three weeks. He was then twenty-three months old.

A boy one year and a half old heard some one say one day that another boy had fallen and hurt his leg. Some days after, the second boy came in, whereupon the first ran up to him, exclaiming, "Fall, hurt leg." A child of two years, whose mother had made him a toy sled out of a card, on receiving a postal card at the door some days after, ran with it to his mother, crying, "Mama, litten" (Schlitten, sled) [72].

New experiences call up memories of old experiences by association, and in this way events that occurred prior to the period of learning to speak, are remembered after that time. A little boy of my acquaintance related the following tale, the events of which took place before he learned to speak: "Pussy kime on table; pull Nonie off (*i.e.*, Nonie pulled her off); pussy katch Nonie face, hands too." This was illustrated by gestures, showing the process of scratching [31]. Another boy, three years old, remembered perfectly well and would imitate his own awkward attempts at speaking [72].

A very interesting question in this connection is this: Which of the senses furnishes the most vivid and lasting memory-images? The first impulse would probably be to attribute the preëminence to sight, but in so doing, we might make a mistake. It is probable, as M. Queyrat seems to think, that the muscular sense is of paramount importance here. Children are full of *action*, and their psychic life is bound up with movement. If they are to develop, they must *do* something, and they remember what they do, a thousand times better than what is told or shown to them. This is also true in adult life. Many persons study out loud. We remember what we *write*, better than what we simply *read*. Pedagogy is now recognizing this as a great principle in education, and the whole kindergarten system is based upon it.

In connection with hearing, the child remembers best some connected story which is helped out by gestures appealing to the eye. The little boy C., at twenty-five months, reproduced after his own fashion the story of Little Red Riding Hood (having heard it only once, and that the night before) with abundant gesture, and then laughed in great glee.

An interesting experiment in this direction is reported by Baldwin in *Science* for May 2nd, 1890. The child was six and a half months old. Her nurse had been absent three weeks. On returning she first *appeared* before the child without speaking, then she *spoke* without appearing. In neither case was she recognized. But when she appeared again, and sang a familiar nursery rhyme, the child recognized her with demonstrations of joy. This is a good example of the "summation of stimuli," or the coöperation of different sensations, reinforcing each other, to produce a result which neither could accomplish by itself.

III. Association.

Memory and imagination proceed in accordance with the laws of association. The chief of these are resemblance, contiguity and contrast. The general principle of association has been expressed in this way: "When, for any reason, a part of an old mental movement is reinstated, there is a tendency for the whole movement to reinstate itself" [Y]. The physiological under-structure of association scarcely exists at birth, but gradually, through experience, dynamic pathways in the cerebral substance are developed, constituting an associative network, connecting the various centres with one another. On the mental side an increasing readiness to note resemblances, differences, etc., and to note them where they are less obvious, is developed in the course of experience.

In Mr. Darwin's opinion, the child far surpasses the lower animals in associative power. "The facility with which associated ideas . . . were acquired, seemed to me by far the most strongly marked of all the distinctions between the mind of an infant, and that of the cleverest full-grown dog I ever saw" [20].

The recorded observations on this point show great individual differences. Champneys saw signs of association of pleasurable feelings as early as the eighth week, when the child accompanied a smiling expression with sucking motions of the lips. Tiedemann thought he saw traces of association on the eighteenth day, when the child ceased crying and put himself into the attitude for taking nourishment when a soft hand came into contact with his face. Sully observed a similar thing at ten weeks. Darwin, on the contrary, did not notice any signs of associations firmly fixed before the fifth month; and Taine puts it as late as the tenth month; while Perez believes that homogeneous sensations are, by the middle of the first month, associated to such a point that they are recognized when reproduced; and he goes on to say that "there is not one of the combinations of associations, which have been studied so carefully by psychologists, of which we cannot find at least a faint foreshadowing in a child of six or seven months" [66].

The following are examples of association by contiguity: When a little child's hat and cloak are put on, or he is placed in his carriage, he becomes restless, and even angry, if not immediately taken out. This has been observed in children less than half a year old [21], and in others of one year [101]. At the latter age the association is much stronger; he cannot even *see* a hat, cloak or umbrella without manifesting the same restlessness. Probably also, as Perez thinks, we may see in the child's crying for food on the return of daylight the germ of association by succession, out of which is constructed the idea of time. A rudimentary notion of cause and effect may also be seen in the babe of half a year or thereabouts, who, having been once burnt by a hot object, afterwards draws back at the sight of it [66]; and in the child, who, finding a peculiar scratching sound to follow the passage of his finger nail over an object, repeats the process again and again, until he has clearly established the relation between the motion and the sound [100]. Contiguity in the form of coexistence is seen in the following: At seven months, the person of the nurse was associated with the sound of her name; when her name was uttered, the child would turn round and look for her [21]. The same thing was observed in another child five months old [66].

Darwin's boy, at nine months, associated his own name with his image in the mirror. When ten months old he learned that an object which caused a shadow to fall on the wall *in front* of him, was to be looked for *behind*. When less than a year old, it was sufficient to repeat a short sentence two or three times at intervals, to fix firmly in his mind some associated idea.

Resemblance, if not the earliest, is certainly among the strongest of the child's associations. Darwin's child, in the second half of his first year, would shake his head and say *ah* to the coal-box, to water spilt on the floor, and to such things as bore a resemblance to things which he had been taught to consider dirty. Another boy, nine months old, on hearing the word "papa," would hold out his arms to another gentleman who resembled his father [66]; and a little girl of this age knew the portrait of her grandfather as it hung on the wall. Sigismund says: "I showed my boy — not yet one year old — a stuffed woodcock, and said 'Vogel.' He immediately turned his eyes to another part of the room, and looked at a stuffed owl which stood there." Taine's little girl, at fifteen months, on seeing colored pictures of birds, immediately cried out *koko*, which was her name for chicken. The little boy, C., on seeing the image on a postal card, at once made a peculiar snuffing noise, which his grandfather was in the habit of doing, showing that he observed a resemblance between his grandfather and the picture on the card.

For resemblances among sounds, children in general have the keenest relish. They are inveterate punsters. Rhymes and alliterations are their especial delight. They will catch the faintest link of resemblance in the sounds of words. "*Harry O'Neil* is nicknamed *Harry Oatmeal*, . . . *October* suggests *knocked over*, and from *do re me*, they get *do re you*" [35]. Mere jingles, tiresome to the grown-up person, will amuse them for hours; such as "Ene, mene, mine mo," etc., or, "Dickory, dickory, dock," etc.

When the child learns to speak, the power of association by resemblances, in his mind, is exemplified in his habit of enlarging the denotation of words, so as to make one word do duty for several objects which resemble each other in certain respects. The discussion of this will be resumed later (*infra*, Section 5 and Chap. V.).

IV. IMAGINATION.

There are two species of imagination. First, the *passive*, in which, without the exercise of active attention, or any effort of will, images pass and repass, arranging and rearranging themselves in the phantasy. This is exemplified in dreams, and in the resuscitation of faded memory images in the waking moments by the laws of association. Secondly, the *active* or *constructive* imagination, in which, by an effort of attention and will, old images are worked up into new forms, inanimate objects have life and personality attributed to them, and curious scenes and combinations are produced by the inventive genius of the person imagining.

With regard to the first, Perez says: "The child, hardly a month old, who recognizes his mother's breast at a very short distance, shows, by the strong desire he has to get to it, that this sight has made an impression on him, and that this image must be deeply engraven on his memory. The child who, at the age of three months, turns sharply round on hearing a bird sing, or on hearing the name coco pronounced, and looks about for the bird cage, has formed a very vivid idea of the bird and the cage. When, a little later, on seeing his nurse take her cloak, or his mother wave her umbrella, he shows signs of joy, and pictures to himself a walk out of doors, he is again performing a feat of imagination. In like manner, when, at the age of seven or eight months, having been deceived by receiving a piece of bread instead of cake, on finding out the trick, he throws the bread away angrily, we feel sure that the image of the cake must be very clearly imprinted on his mind. Finally, when he begins to babble the word *papa* at the sight of any man whatever, it must be that the general characteristics which make up what he calls *papa* are well fixed in his imagination." When they are left alone, children who have acquired the word "mamma," will repeat this name over and over again, proving the presence of the mother's image in the imagination [66].

One of the most significant forms of the passive imagination in childhood is the dream. It is very difficult to ascertain when the child first begins to dream, and this for several reasons. The child who can talk, will "tell his dreams," in imitation of grown-up people, no dream having taken place. In the case of the child who cannot talk, we have very little reliable information to go upon. But there seems no reason to doubt that dreams may take place just as soon as the child's waking experiences have furnished him with clear and definite sensations.

As for the *constructive* imagination, our space will not admit the hosts of examples that might be given of the wonderful fertility of children's minds in this respect. Their little wooden toys become transformed into real soldiers, fighting real battles, mighty locomotives drawing long trains of heavily-laden cars, or great steamships sailing over unfathomable oceans. "Given a few broken pieces of glass, a flower, a fruit, a colored string, a doll, and out of them the baby imagination constructs an immeasurable happiness" [57].[1] Indeed it would seem, as Jastrow says, that the

---

[1] See "The Story of a Sand Pile," by G. S. Hall, in *Scribner's Magazine* for June, 1888.

function of toys is to serve as "lay figures, on which the child's imagination can weave and drape its fancies" [41]. In order to serve this purpose, the toy does not need to be a work of art. "We don't like buyed dolls," says little Budge, in "Helen's Babies," and in so saying, he seems to voice the opinions of the majority of children. A wax doll is a nice thing to have, and look at occasionally, but for real, "sure enough," every-day play, give us the old rag doll.[1]

Children in their plays imagine themselves other than they are. They transform themselves into kings and queens, professors and preachers, fathers and mothers and grandparents, and fulfill all the functions of neighbors and citizens with the greatest solemnity and dignity. They surround themselves with imaginary personages, and carry on imaginary conversations.[2]

I shall close this section with a quotation. W. W. Newell, in "Games and Songs of American Children," says: "Observe a little girl who has attended her mother for an airing in some city park. The older person, quietly seated beside the footpath, is half absorbed in reverie; takes little notice of passers-by, or of neighboring sights or sounds, further than to cast an occasional glance, which may inform her of

---

[1] The same thing holds with regard to pictures. I have seen a copy of a German picture-book for children, which is almost completely lacking in artistic excellence, but which has gone through one hundred and seventy-seven editions. A movement is now on foot in Russia to prohibit the importation of the finely finished and elegant French toys, on the ground that they leave no room for the exercise of the child's imagination.

[2] "One of the greatest pleasures of childhood is found in the mysteries which it hides from the scepticism of the elders, and works up into small mythologies of its own." Holmes, "The Poet at the Breakfast Table."

the child's security. The other, left to her own devices, wanders contented within the limited scope, incessantly prattling to herself; now climbing an adjoining rock, now flitting like a bird from one side of the pathway to the other. Listen to her monologue, flowing as incessantly and musically as the bubbling of a spring; if you can catch enough to follow her thought, you will find a perpetual romance unfolding itself in her mind. Imaginary persons accompany her footsteps; the properties of a childish theatre exist in her fancy; she sustains a conversation in three or four characters. The roughness of the ground, the hasty passage of a squirrel, the chirping of a sparrow, are occasions sufficient to suggest an exchange of impressions between the unreal figures with which her world is peopled. If she ascends, not without a stumble, the artificial rockwork, it is with the expressed solicitude of a mother who guides an infant by the edge of a precipice; if she raises her glance to the waving green overhead, it is with the cry of pleasure exchanged by playmates who trip from home on a sunshiny day. The older person is confined within the barriers of memory and experience, the younger breathes the free air of creative fancy."

V. The Discursive Processes.

Conception, judgment and reasoning, the three processes of discursive thought, are treated together, because it is impossible to make qualitative distinctions among them. They differ only in degree, not in kind. In every concept, there is involved a rudimentary judgment, and the syllogism consists simply in the apperceptive synthesis of judgments, whose constituent elements are concepts. The three are then at bottom only different stages in the one process, by which knowledge of the abstract is elaborated. Examples given, therefore, to illustrate the one, contain elements almost equally illustrative of the others.

Conception. — The child's earliest experience, being predominantly physiological, is also predominantly individual and concrete. He lives in the particular. It is a momentous juncture in his life when he first steps out beyond individual things, to abstract their common qualities, and of these to form notions. It is only then that he begins to *think*, in the strict sense of the word; and it is this thinking in abstractions and generals, which, in Locke's opinion, differentiates the human mind essentially from lower animal intelligence.[1]

Taine believes that the general notion makes its appearance only with the acquisition of language. Preyer, on the other hand, maintains that "even before the first attempts at speaking, a generalizing and, therefore, concept-forming combination of memory-images regularly takes place." "That the ability to abstract may show itself, though imperfectly, even in the first year, is, according to my observations, certain. Infants are struck by a quality of an object — *e.g.*, the white appearance of milk. The 'abstracting,' then, consists in the isolating of this quality from innumerable other sight-impressions, and the blending of the impressions into a concept. The naming of this, which begins months later, . . . is an outward sign of this abstraction, which did not at all lead to the formation of the concept, but followed it" [72]. He also quotes from Oehlwein to show that deaf-mute children, in the first year of life, form concepts, and logically combine them with one another; and he concludes that thinking is not bound up with verbal language, though it no doubt demands a certain

---

[1] "Human Understanding," Book II., Chap. II.

degree of cerebral development. Even orangs and chimpanzees reason without language, but their concepts are neither so abstract, so clear, nor so numerous as those of the child even before he learns to speak, while after that time the gulf between them widens infinitely.

Perez agrees with the above view, and quotes from Houzeau to show that dogs, bees and other dumb creatures have concepts, and carry on reasoning processes. As to the child, he gives several examples on this point. A boy of eight months, who used to amuse himself by stuffing things into a tin box, afterwards examined every new toy to find an opening. Another child of the same age used to make a peculiar sound when he desired solid food, different from that by which he expressed his desire of the breast. Another, at nine months, gave unmistakable evidence that he possessed the concept "animal."

According to Romanes, there is a class of ideas standing between the percept and the concept, less abstract than the latter, but more general than the former, to which he gives the name *recept*. They are complex ideas arising out of a repetition of more or less similar percepts. *E.g.*, when a parrot, who has learned to call out *bow-wow* when the house dog enters the room, also calls out this word on seeing other dogs of various sizes, colors and forms, he possesses an idea which constitutes an advance on the percept, but cannot, strictly speaking, be called a concept. Every child passes through a *receptual* stage, which does not require language, whereas the concept, properly so-called, or the active synthesis of qualities into a class is not, in his opinion, attained until the child can speak.[1]

---

[1] See also a series of articles in *Public School Journal* for November and December, 1891, and January and February, 1892, entitled, "How do Concepts arise from Percepts?"

Taking the ordinary meaning of the word *concept*, which includes what Romanes expresses by *recept*, it seems established that the formation of the concept is prior to, and in large measure independent of, language; but it seems equally clear that abstraction and generalization do not attain to any great degree of complexity without the aid of speech, as the observation of the cleverest deaf-mutes clearly shows. Even after speech begins, the discursive processes develop but slowly. In one case, a child of seventeen months had not yet differentiated his collective concept "taste-smell" (as united in one object) into the concepts "taste" and "smell" [72]; though another child, at seven months, seemed to have ideas of kind [66]. A boy of three years did not know the meaning of "size" or "goodness," though long before this he perfectly understood the expression: "Baby is a good boy." Children have very little idea of number in the first two years. A child of two and a half years confounded "naughty" with "ugly." In short, we find at this period only the lowest degree of abstraction.

The child's first generalizations are very inaccurate. Even when he begins to talk and to use general names, he does not use them in the same sense as the adult. His generalizations are apt to be too wide. "Logic in the child naturally operates with much more extensive and less intensive notions than in adults. Hence he is very liable to illusion, not through stupidity, but simply through ignorance, arising out of lack of experience." After having held out grass to a sheep, he also offers some to the birds [72], and in this he is acting with perfect consistency, within the range of his knowledge. He extends the term *papa* to other men, the word *atta* or *peudu* (perdu) to all sorts of disappearances; he makes the word *quack-quack* apply not only to a duck, but to the water on which the duck swims, then to all birds and insects, then to all fluids, and finally to all coins.

because he had seen the picture of an eagle on a French sou [81]. He includes an eye-glass in the concept *bon dieu* (blessed medal), and the steamboat, coffee-pot, and all hissing, noisy objects, in the class *fafer* (chemin de fer. locomotive). A little girl of eighteen months had been amused by her mother hiding in play, and saying *coucou*. She had also been warned to keep out of the hot sun, by the words *ça brule*. One day, on seeing the sun disappear behind a hill. she put these two ideas together and exclaimed *a bûle coucou* [99]. Another child of the same age applied the name *no-no* to all eye-glasses, because she had been forbidden to snatch off her nurse's glasses by the words *no-no* [B]. Taine believes the characteristic mark, distinguishing the child from the lower animal, is this very capacity of detecting resemblances amid differences, which leads him to extend, to such a surprising degree, the denotation of the term. Not only does he apply the word *bow-wow* to the terriers, mastiffs and Newfoundlands which he meets in the street, but "a little later he does what an animal never does, he says *bow-wow* to a pasteboard dog that barks when squeezed. then to a pasteboard dog which does not bark, but runs on wheels, then to the bronze dogs which ornament the drawing-room, then to his little cousin, who runs about the room on all fours, then, at last, to a picture representing a dog [99].

Children's notions of things are chiefly connected with their uses or actions. M. Binet gives a large number of interesting definitions of things given by children, from which I select the following: " Un couteau, c'est pour couper la viande." " Un cheval, c'est pour trâiner une voiture, avec un monsieur dedans." " Une lampe, c'est pour allumer, pour qu'on voie clair dans la chambre." " Un crayon, c'est pour ecrire." " Un chapeau, c'est pour mettre sur la tête." (Note the frequency of the " pour.")

JUDGMENT is involved, in a rudimentary form, in conception, and even in perception, as may be seen from the foregoing examples. When a child at two months recognizes his parents; at three and a half months turns round to the cage on hearing the word *coco;* "comes to meet" the spoon with his mouth when being fed; at seven months turns his head around to the left when an object is carried so far behind him that he can no longer see it by turning to the right; at eight months recognizes a pictorial representation; and cries for Gourlay water, which is white and opaque, though not for ordinary water; in the tenth month gives evidence of the knowledge that bodies have weight; and shows by unmistakable signs that he misses his absent parents, and even knows when a single nine-pin is removed from his set, — we cannot doubt that he is performing an act of judgment. These primitive judgments are mostly concrete and particular, abstract and general judgments being a later attainment. Children of eighteen months will recognize the pictures of all the more familiar animals, and respond with the appropriate sounds, *bow-wow, moo,* etc. The spoken judgment arises when an object arouses an idea in the child's mind, to which idea he attaches a name, recognizing it as connected with the object. The first spoken judgment does not then require two words, as Taine seems to think, but usually consists of one word, which does duty for a whole sentence.[1]

REASONING. — When the little boy, R., was four months old, he was playing one day on the floor surrounded by his toys. One toy rolled away beyond his reach. He seized a clothes-pin and used that as a "rake" with which to draw

---

[1] Preyer's boy, at twenty-three months, uttered his first spoken judgment, viz., "Heiss" (= "This food is too hot").

the toy within reach of his hand. Mr. Darwin laid his finger on the palm of a child five months old. The child closed his fingers around it, and carried it to his mouth. When he found that he was hindered from sucking it, by his own fingers getting in the way, he loosened his grasp and took a new hold farther down, then vigorously sucked the finger. When Preyer's boy, at six months, "after considerable experience in nursing, discovered that the flow of milk was less abundant, he used to place his hand hard upon the breast, as if he wanted to force out the milk by pressure." Another child, at seven months, cried for a share of the food his nurse was eating [66]. A boy of eight months took a watch, which was offered him, and after biting on it with evident satisfaction, tried to break a piece off, as he would from a cracker. At thirteen months, a child who noticed the resemblance between two men, inferred certain acts on the part of the one, which he had been accustomed to see in the other [11].

The boy, C., when fourteen months old, was one day feeding the dog with crackers, when the supply ran out. He immediately "crept to the sideboard, opened the left-hand door, pulled himself up by the shelf, and helped himself out of the box in which they were kept." He had seen crackers taken from this box before, but had never done it himself. He was observed to feel his own ears, and then his mother's, one day when looking at pictures of rabbits. One day, when eighteen months old, he came in from playing on the lawn, quite hot and somewhat dirty. He at once ran to his mother, holding up his dirty dress with a gesture of disgust; then ran to the drawer where his clean clothes were kept, and tugged at it with all his might. Another boy of the same age, both of whose hands were filled with toys, wishing to grasp still another, quickly put one of them between his knees [104]. A little girl of this age used to feign sleep until the nurse left the room, when she would immediately resume her interrupted romps ^(B). Tiedemann's boy, at two years of age, used to employ cunning to accomplish his purposes. The little girl. F., at a year and a half, furnished a good example of reasoning by analogy. She had been shown the pictures in a book with red binding. She afterwards went to the bookcase and took down two other books having red binding, and looked through them, evidently expecting to find pictures in them also. One day when I rose to take my leave, she patted vigorously on the cushion of a chair, and then pulled at my coat to induce me to prolong my stay.

From about the end of the second year, the reasoning power in most children makes such rapid progress that it is impossible to set down all the examples that are to hand. I content myself with one more. A boy of two years was quite familiar with the articles of his food by name, and when the word *milk* was spoken in his hearing, he clamored for a share of that article. His mother hit upon the device of spelling the word, when it was undesirable that his attention should be called to it. Before long, however, he learned to know the word, even when spelled. and one day, when his mother asked for the m-i-l-k, he at once cried out, *mickey* [A].

VI. The Idea of Self.

The phenomena which accompany and indicate the gradual emergence into clear consciousness, of what Taine calls the "unextended centre," the "mathematical point," by relation to which all the "other" is defined, and which each of us calls "I." or "me."—the external evidences that the child is slowly but surely becoming "aware of himself as a permanent being. distinct from the objects he knows, the feelings he experiences, and the ends he chooses" [Y]. — may be conveniently classified under four heads:

1. THE CHILD'S TREATMENT OF HIS OWN BODY. — In the first weeks he will strike or scratch his own face. One boy bit his own finger until he cried with the pain, even in the early part of the second year. In the ninth month the feet are still eagerly felt of, and the toes carried to the mouth, the same as foreign substances. This experimentation with his own limbs goes on all through the second, and in some cases well on into the third year. "In the first year the child's organism is not known as part of himself" [81]. A boy of nineteen months, when asked to "give the foot," seized it with both hands, and tried to hand it over [72]. A little girl, a little over two years old, used to enlarge on a familiar ditty in the following fashion: "One for papa, one for mamma, one for toses (one for toes)" [B]. Sigismund believes that the child learns a good deal about his own limbs (and so takes the first step toward a knowledge of self) through bringing his hand to his mouth, to ease the pain of the growing teeth. The feeling is different when he chews his own finger and that of his nurse. A child of four or five months studies his own fingers attentively. When one hand accidentally grasps the other, he looks attentively at both. Lying on his back, he gazes at his legs stretched up in the air.

Closely connected with this is the child's evident delight in his own activity and ability to do things. Wundt believes the muscular sense plays a predominant *rôle* in the genesis of self-consciousness, and there is little doubt that the acquisition of the power of walking contributes very largely to the growth of the self-idea. The feeling of power is engendered by the discovery that he can cause changes in objects. "An extremely significant day in the life of the infant is the one in which he first experiences the connection of a movement executed by himself with a sense-impression following upon it." [72]. Preyer's boy, in the fifth month, discovered that by tearing paper he could produce sound sensations; also by shaking a bunch of keys, opening and closing a box (thirteenth month), turning the leaves of a book, etc., and these occupations were accordingly carried on with a perseverance astonishing to an adult. He experienced a genuine pleasure in finding himself a cause.

2. THE CHILD'S BEHAVIOR TOWARDS HIS IMAGE IN THE MIRROR. — Darwin's child failed to interpret his reflection when five months old, but two months later he had accomplished it, and at nine months had learned to associate his name with the image. Another child at eight months used to look at his reflection with wonder (expressed by wide-open eyes and immobility). " On being shown a hand glass, he regards his image with interest, smiles and tries to catch it. He puts his hand on the glass, and tries to take hold of the image's hand. Then he turns the glass over, and looks up in wonder at the result " [11]. A similar performance was gone through by a boy of ten months; and, six months later, he was found one day standing before the glass, pulling his hair, examining his eyes and ears, and sticking out his tongue [C]. Preyer's boy did not notice himself in the glass when three months old. Three weeks later he looked at it, but with indifference. Two weeks later still, he regarded it with attention, and laughed at the sight of it. Near the end of the sixth month, he stretched out his hand towards it. In his ninth month he grasped at it, and seemed surprised when his hand came against the smooth surface. At fourteen months he passed his hand behind the glass, as if searching for something. He afterwards behaved in the same manner toward a photograph.

In the sixteenth month he made grimaces before the glass, laughing as he did so. Two weeks later he looked at himself often in the glass, with pleasure and evident vanity. At twenty months he connected his own name with the image, and when asked, "Where is Axel?" would point to the reflection. Another child knew her image in the glass at twelve months, would point to it and say *Tatie* (Katie). A little boy of fifteen months calls his image *Titta*, by which he means *child* or *doll*.

3. In the third place, we have those actions which show the beginnings of THE FEELING OF PROPERTY, such as pride in personal appearance, and in adornment, jealousy over toys, and other things which the child considers his rights. A number of examples have already been given in connection with the emotion of jealousy. As regards personal adornment, there are very great differences among children, some taking great delight in it, while others seem to care but little about it. A little girl whom I have observed since her first year seems very fond of it, and will spend hours in adorning herself with veils and feathers and bracelets, making believe she is some fine lady. Whenever her best clothes are put on, or a new hat, she is very proud and walks very straight and dignified indeed.

4. Lastly, we notice THE CHILD'S USE OF THE PRONOUN "I" (*Je, Ich, Ego*). It is interesting to remember that, according to the opinion of some philologists (Max Müller, for example), this word was, at the beginning of the development of language, a demonstrative, meaning "this one," and was probably accompanied by a gesture, and perhaps, further back still, the gesture supplied the place of the word. Man spoke of himself in the third person before he learned to use the first person. Just so with the child. He first calls

himself by his proper name, or he uses the word *baby*, and the intelligent use of the first personal pronoun comes late — most observers put it as late as the third year. I have never heard a child less than two years old call himself "I" or "me." The chief difficulty in the way of his doing so is that he never hears the word applied to *him* by others. This is why he makes such errors as "Take me up on *my* (meaning *your*) lap."

The "I" feeling is often present, therefore, before the word is used. The concept of the self is not *generated*, but only rendered more exact and definite by speech. On the other hand, it must not be presumed that the concept is always present where the word is used. Children who are constantly in the society of those who use the word will use it also, merely by imitation in many cases, without comprehending its meaning. A child may say "I am hungry," without any idea that "I" is different from "hungry" [81]. Perez says: "When the child learns to say 'I' or 'me,' instead of 'Charles' or 'Paul,' the terms 'I' and 'me' are not more abstract to him than the proper names which he has been taught to replace by 'I' or 'me.' Both the pronouns and the names equally express a very distinct and very concrete idea of individual personality. When a three-year-old child says 'I want that,' it is only a translation of 'Paul wants that,' and 'I,' like 'Paul,' indicates neither the first nor the third person, but the person who is *himself*, his own well-known personality, which he continually feels in his emotions and actions. An abstract notion of personality does not exist in a young child's mind" [66]. In short, so great is the influence of the environment here, that scarcely anything can be asserted in a general way of all children. Some children scarcely ever hear the pronoun "I." The members of the family avoid it, and say instead: "Mamma is busy;" "Sister is gone to school;" "Baby must be good," etc.; in such cases, the child will of course take a long time to acquire the word.

In many cases, *me* is used before *I*. It seems easier, for some reason. Sometimes children pass through a sort of transition period, when *I* is used indifferently with the proper name, or even with *he*. Binet says of the little girl he observed that at three and a half years exactly, she first used the word *je*, in the sentence *je ne sais pas*. Two days after she said *je ne veux pas*. But long after that, she made many mistakes in the use of the pronoun. In two other children, the *I* took the place of the third personal designation before the end of the third year, and *I* preceded *me*, and *you* was later than either [36]. Another child at twenty-five months used *my*, but not *I* [B].

Such are the various factors entering into the development of the child's self-consciousness, by which "he raises himself higher and higher above the dependent condition of the animal, so that at last the difference (not recognizable at all before birth, and hardly recognizable at the beginning after birth) between animal and human being" attains such infinite magnitude.

CHAPTER IV.

VOLITION

We now approach the most difficult as well as the most important part of our subject: the most difficult, because of the exceedingly complicated character of every act of will; the most important, because of the vast influence which any one's theory of volition must exert upon his moral and religious ideas. Not only is it true that "a being is capable of education and morality in proportion as he is capable of will" [33], but it is also true that the most widely separated views touching human responsibility and destiny, have grown out of apparently slight differences of opinion with regard to the nature and freedom of the will. The following theories are quoted to show the trend of contemporary opinion on the subject, and not to set forth the present writer's views.

"Out of the desire of everything that has once occasioned pleasurable feelings, is gradually developed the child's will" [72]. In Preyer's view, the will is called into life by the union of two representations, viz.: 1st, that of the end desired; 2nd, that of the movement necessary to attain the end. The latter is not absolutely necessary, and at a later period is no longer formed, except in the case of new movements. The idea of the end is sufficient, without that of the means. Will, then, is based upon, and grows out of, desire.[1]

---

[1] Preyer's theory of the *origin* of will is not, however, an empirical one, as the following quotation will show: "It is an error to think

In Guyau's opinion, also, a complete act of will involves representations of two sorts, viz.: Of the act about to be performed, and of another, contrary act, which *might* have been performed. Action, then, is the resultant of a struggle among tendencies.[1]

Perez says: "The will is born little by little from reflex, impulsive and instinctive movements, which, with the progress of the faculties of perception and ideation, and after having been for a long time executed and varied, fall under the action (*coup*) of the attention, and become conscious, reflected, and, in a word, voluntary." Will in its negative form (inhibition), he holds to be also a matter at first of mechanism, unconscious and involuntary. It is a suppression, or at least a reduction, of reflex, impulsive and instinctive movements, by the fact of an excitation of the brain, a sensation. Thus arrest consists at first simply in the substitution of one tendency for another.[2]

Wundt, on the contrary, holds that there is no such thing

---

that the will arises from impressions in youth; . . . a will can never be created in a child from external experiences; it must be allowed to develop itself from the inborn germ of will" (74).

[1] "La pleine volonté, c'est-à-dire le déploiement total des énergies intérieures, suppose qu'à la représentation de l'acte même qu'on va accomplir, s'associe la représentation affaiblie de l'acte contraire. Et ainsi, nous arrivons à cette conclusion: Il n'y a pas d'acte pleinement volontaire ou, ce que revient au même, pleinement conscient, qui ne soit accompagné du sentiment de la victoire de certaines tendances intérieures sur d'autres, conséquemment d'une lutte possible *entre* ces tendances, conséquemment enfin d'une lutte possible *contre* ces tendances" (32).

[2] See also Ribot, "Les Maladies de la Volonté," p. 8. Bain, "The Emotions and the Will," Part II. Chap. I., and compare Baldwin's "Deliberative Suggestion," in which various "coördinated stimuli meet, affront, oppose, further one another, . . . response answering to appeal in a complex but yet mechanical way" (7).

as purely reflex and involuntary consciousness; that activity of attention is in some degree present even in movements apparently the most mechanical.[1]

Professor James lays down, as the distinguishing mark of voluntary movements, an antecedent *desire* and *intention* to perform, and consequently a full *prevision* of what the action is to be. He therefore designates voluntary movements as secondary functions of our organism, while "reflex, instinctive and emotional movements are all primary performances." He makes voluntary movements depend on memory-images of former involuntary ones. "When a particular movement, having once occurred in a random, reflex or involuntary way, has left an image of itself in the memory, then the movement can be desired again, proposed as an end, and deliberately willed. But it is impossible to see how it could be willed before. *A supply of ideas of the various movements that are possible, left in the memory by experiences of their involuntary performance, is thus the first prerequisite of the voluntary life.*"

It will be seen that all these views corroborate the position taken in the present work, that mental phenomena undergo a process of transformation, in virtue of which, from being predominantly physiological, they become predominantly psychical. We see now the application of this law to movements or actions. The earliest child movements, in the opinion of these writers, are not voluntary, but only reflex, instinctive, etc. Intelligent apprehension of the end sought, and of the means by which that end is to be attained, has not yet taken place, and, we may add that, until it has taken place, the *movement* is no more entitled to be called an *action* than is the swaying of a branch in the breeze, or the "action" of the piston-shaft of a locomotive. The conscious

---

[1] "Menschen und Thierseele."

subject must first take hold of the *movement*, and put himself forth in intelligent direction of that movement toward a conceived and desired end, and then it becomes *transformed* into an *action*. It seems necessary also, in order to avoid misunderstanding, to express our dissent from the view held by some of these writers, that the will is a derived product, or result of mechanical movements, a something which has been brought to the birth by the "travail together" of accidental motions in an animal organism. It is an obvious *hysteron proteron* to explain the rise of will by means of this principle of transformation, while the only possible way of explaining the transformation is by positing voluntary activity. It is said, for example, that will is *born* (!) little by little out of reflex and instinctive movements, which have come within the scope of the attention; and again that will is developed out of the desire of everything that has occasioned pleasurable feeling. Now both attention and desire, as we understand them, are impossible without volition. They involve active direction of the self toward the object, and this is volition. So far, then, from being the antecedents of will, they are modes of its manifestation, and instead of ascribing the birth of will to the transformation already spoken of, in virtue of which movements come within the scope of the attention, we should more correctly ascribe the transformation to the exercise of will. The will is the cause and not the effect of the transformation. It is correct enough to say with Preyer that will is *developed* in connection with these movements and desires — if by *development* is meant only *growth* and not *genesis* — but when it is asserted that will is *generated* out of actions to which attention and desire are directed, it is only necessary to ask: Out of what are attention and desire generated? to reveal at once the insufficiency of the explanation.

This criticism is all the more necessary here, because

Professor Preyer's classification of child-movements,— as the most scientific and exhaustive yet made,— is adopted in the following pages. It can be accepted in toto, *as a description and classification* without our subscribing in the least to any particular theory of will-genesis that may have been founded upon it. The classification is as follows. First, we have a multitude of movements, not involving peripheral stimuli, but proceeding entirely from internal conditions. They are simply the result of an overflow of nervous energy, and require only motor — not sensori-motor — processes. They are, of course, will-less, and are designated *impulsive* movements. Secondly, we have those movements (very numerous in the new-born) which, though requiring peripheral stimuli, and, therefore, sensori-motor processes, do not involve active attention or effort, and are, therefore, will-less. These are the well-known sensori-motor *reflexes*. In the third place, there is a kind of movements — found in great abundance in the human being, and constituting, probably, the majority of the so-called actions of the lower animals — for which the physical and emotional organism is specially fitted by the action of heredity. These are the *instinctive* movements. Finally there supervene on all these the *bona fide actions* of the person, involving desire of end, attention to the object, and representation of, and deliberation upon, the means of attainment, as well as the conscious forth-putting of the self in *effort* towards the realization of the represented end. These are the *ideational*, or consciously deliberated and voluntary movements. We shall consider these in this order, only premising that because any given movement is here classed as impulsive or reflexive, it does not necessarily follow that it is never to be found in any other class. A movement, the same outwardly, may be at one time impulsive and at another ideational. This is one application of the principle of transformation.

I. Impulsive Movements.

The majority of the embyronic movements belong to this class. From the time of "quickening," the fœtus performs numerous muscular movements (mostly set on by processes of nutrition and circulation) prior to the first exercise of reflex sensibility. In the new-born they are still numerous, comprising all those spontaneous kickings and rollings, awkward muscle-movements and comical grimaces, so noticeable in the early weeks of life. The hands strike right and left and move toward the face without any definite object; the legs tramp and kick when the child is held up in the air; the eyes may be observed to move before the lids are opened; the intra-uterine posture is resumed on falling asleep; the limbs are stretched on awakening; in short, almost every muscle of the body is exercised without any assignable peripheral stimulus. The movements are often symmetrical (by accident), but usually at first asymmetrical. Some of them (as yawning and stretching) persist through life, but the majority have disappeared by the end of the second year. Many of them are unexpected by the child himself; he is evidently surprised to find himself performing a certain movement, and afterwards performs it voluntarily, with numberless repetitions, and evident pride in the newly discovered ability.

The first smile doubtless belongs here, as also the peculiar crowing heard so frequently in the first year; and the numerous "accompanying" movements made by the child (such as holding the hands in a certain strained position, with the fingers spread out, while drinking, and the dreamy, wandering motions of the eyes during the act of sucking). A sleeping child suddenly threw up one of his hands, which, coming into contact with the eye, pushed the lid open. The infant slept on with one eye open,— the pupil very much contracted — until by-and-by the hand dropped and the eye closed [72].

Although possessing in themselves no direct volitional significance, yet these impulsive movements are indirectly of great importance, inasmuch as they are the raw materials, upon which the gradually awakening child-will exercises itself, making them its own, and transforming them, by means of conscious activity, into voluntary *actions* properly so-called.

II. REFLEX MOVEMENTS.

These occur as the response of the nervous system to peripheral stimulation, without the participation of the idea. If they enter into consciousness at all, it is only during or after their performance. They are found in the adult in great abundance as well as in the child; and are very well exemplified in the sudden movements of the hands when one's hat is blown off in the street. Though heredity probably plays a considerable part in facilitating them, yet they do not take place in the earliest infancy with that certainty and promptness by which they are characterized in later life, as we have seen in the case of eye movements. What seems to be transmitted is a potentiality, which needs experience to transform it into an actuality.

The law of transformation has an obvious application here. Indeed we see in the case of these movements a double transformation: that which was at first a reflex movement becomes afterwards a voluntary one; and finally, by virtue of repetition, leading to the formation of a habit, it becomes once more reflex or automatic. Probably all mouth movements involved in the enunciation of articulate sounds, pass through all these stages, as we shall see later.

Reflex movements are of great importance in will-growth,

since upon them the voluntary movements, properly so-called, supervene. On its negative side also (*i.e.*, in inhibition) the will develops chiefly in connection with the repression of reflexes.

In the earlier stages of fœtal life, according to Preyer, no reflex movements can be elicited, be the stimuli never so strong and varied; and even after there have occurred many movements of an impulsive nature. But reflex excitability increases very rapidly in the later months, even gentle stroking calling forth many movements. Swallowing as a reflex occurs at this time; and fœtal movements can be evoked by changes of temperature. Champneys says the curling up of the toes, and jerking away of the foot when the sole is tickled (which Mr. Darwin observed on the seventh day of life), can be produced *in utero*. Only from the beginning of extra-uterine life, however, does the reflex activity of the nervous system obtain full play. And here the earliest and most prominent are the various respiration reflexes. The first cry is undoubtedly of this character, since brainless children make themselves heard in the first minutes of life as well as normal children.[1] Sneezing, too, which in many new-born children takes the place of crying, is a pure reflex, as it continues to be through life, though the complex coördination of many muscles, by which it is accompanied, is not so complete in the child as in the man. Other reflex movements connected with respiration are *coughing, wheezing, choking, laughing* when tickled, *hiccoughing*, and the like, all of which, with the exception of laughter, may probably be observed in the first week. A striking proof of the reflex sensibility of the respiratory apparatus is seen in the fact that a noise, just loud enough *not*

---

[1] See several cases cited by Taine, "Intelligence," Part I. Book IV. Chap. I.

to awaken the sleeping child, has the effect of increasing the rapidity of the respirations [72].

Starting at any sound or jar, does not occur at the very first, but makes its appearance early. Generally there is silence for a moment after the disturbance, as though the energies were temporarily paralyzed. Champneys observed this starting first in the fourth week, but the child would not start twice at the same noise, unless it was very loud. Children are very susceptible to nervous stimuli, as is evident from the frequency of convulsions in infant life.

Reflex movements of the *limbs* are numerous, prompt and early. On the seventh day Darwin tickled the sole of his child's foot with a piece of paper; the foot was jerked away and the toes curled up. He remarks: "The perfection of these involuntary movements shows that the extreme imperfection of the voluntary ones is not due to the state of the muscles, or of the coördinating centres, but to that of the seat of the will." On the fourth day another child clasped a finger laid in his hand [95]. From the fourteenth day on, tickling the sleeping child's temple was followed by a movement of the hand toward the place, though the hand did not always find the right spot [72]. The left hand did not always respond, in Preyer's experiments, to stimulus applied to the left side, nor the right hand to the right side; but Pflüger found the responses constant in this respect.[1] There seem, indeed, to be two sorts of reflexes: the *inborn* (such as spreading the toes on tickling), which occur from the first hour of life with perfect regularity and accuracy; and the *acquired* reflexes, which are neither prompt nor certain at first, but become so on repetition.

Very important in this connection are the reflex *eye-*

---

[1] So also Baldwin. See "Infants' Movements" in *Science*, Jan. 8, 1892.

movements of the new-born child. The examples given in the first chapter of the responses of the infant eye to impressions of light,— turning towards the light, following a moving light or brightly colored object, etc.,— are mostly examples of reflex movements, as are also those movements of the eyes which follow touch-impressions on the lashes, lids, etc. According to Preyer, there are "six different regular reflex movements from the optic nerve to the motor oculi alone, which appear in the case of light impressions."

Least developed of all in the earliest period are the *pain-reflexes*. The new-born in many cases makes no response whatever to the prick of a pin, as Genzmer has shown. The response takes place, however, when the stimulus is such as to affect a large number of nerve ends at the same time (a slap for example). This tardiness of pain-reflexes in the new-born does not show that he is insensible to pain,— though he is, probably, *less* sensitive than the adult in this respect,— but simply that the nerve connections which make reflex movements possible, are in the case of pain sensations less developed than those of the skin and mucous membrane.

Finally the *inhibition of reflexes*, by which the will of the child develops on its negative side, is very difficult, and therefore a late attainment. In one case it was observed as early as the tenth month, when the child for the first time restrained his excretions [72]; in another, during the first quarter of the second year, when the child checked an impulse to scratch [7]; and in a third, in the fifteenth month. In marked contrast to this is the inhibition of reflexes in the lower animals, where it often takes place before the end of the fœtal period.

III. Instinctive Movements.

These differ from impulsive movements in that they do not occur in the absence of appropriate peripheral stimuli. There is in the child an inborn instinct to seize with the hand, but this movement takes place only when the palm comes into contact with an object. They differ from impulsive movements also in having an *end* or *purpose*, though this end may not be known at the time of their performance.[1] Besides the stimulus, they require a certain emotional condition. The child in a sorrowful frame of mind does not laugh when his toes are tickled. They differ from ideational movements in the absence of a pattern, and of any conscious effort, or previous representation.

One of the strongest instincts in the child is to seize objects and carry them to his mouth. Attempts at this have been observed as early as the fourth day. This propensity to make the mouth the test-organ for all sorts of objects, has been explained by the hypothesis that the lips may have been used in conjunction with the hands in an earlier period of race-progress, much more extensively than at present [95]. The movements of the hands to the mouth may be at first accidental, and then instinctive, as in painful teething. It finally becomes reflex through the formation of habits. The contraposition of the thumb in seizing objects is quite slowly learned (in one case as late as the 12th week). This is in marked contrast to the facility with which young monkeys, less than a week old, oppose the thumb in seizing.

As to the rise of right or left-handedness, Professor Bald-

---

[1] " Instinct is . . . the faculty of acting in such a way as to produce certain ends without foresight of the ends, and without previous education in the performance " [40].

win has made a large number of experiments, whose results may be summarized as follows:

(1) No trace of preference for either hand was discernible so long as there were no violent muscular exertions made. In over 2000 experiments, one hand was preferred as often as the other.

(2) From the sixth to the tenth month, the tendency to use both hands together was about twice as great as the tendency to use either hand alone. (The figures are: Number of experiments, 2187; right hand used alone 585 times, left hand alone 568 times, both hands together 1034 times.)

(3) Right-handedness developed under the pressure of muscular effort. Preference for the right hand in violent efforts in reaching appeared in the seventh and eighth months. Experiments made in the eighth month gave this result: Right hand 74, left 5, both 1. Under the stimulus of bright colors, the right hand was employed 84 times, and the left hand only twice.[1]

Often there is a period of left-handedness in children who afterwards become right-handed. Sigismund believes that most children up into the third year prefer to use both hands together.

Among instinctive mouth movements the earliest and most perfect is *sucking*. Sometimes, however, even this movement is far from perfect at the beginning. Many of the earliest efforts are quite fruitless, owing to failure in coördination. This movement doubtless takes place before birth, since it may be observed from the first moments of life. On its development, Kussmaul remarks to the follow-

---

[1] Professor Baldwin sees, in the fact that preference for the right hand was developed only in connection with muscular effort, an argument in favor of the "innervation" theory. For the opposite opinion see a short article by Professor James in *Science*, 14 Nov., 1890

ing effect: An advance is made on the mere reflexes when the child sucks the finger thrust into his mouth, or the nipple of the breast. Here we have not only sensation, awakening movement, but also feelings of pleasure or displeasure, with answering endeavors and mental representations of the simplest kind. Finally the will learns to regulate these movements in the interests of the individual.

Other instinctive mouth movements are *biting* (which begins about the fourth or fifth month, and supersedes sucking from the tenth month), *chewing* (which is performed with perfect regularity from the fourth month), *grinding the teeth* (which is quite original, and probably practiced by all babes during teething), and *licking* (which is performed in the first twenty-four hours " hardly less adroitly than in the seventh month ") [72].

Learning to *walk* involves a whole series of preliminary accomplishments, first among which is the ability to hold the head in equilibrium, which may be accepted as the criterion of the rise of voluntary power. This is usually accomplished about the fourth month. The next stage is reached a month or two later in the ability to sit alone upright. When this is successfully accomplished for the first time, the soles of the feet are frequently turned towards each other — a partial re-assumption of the intra-uterine posture. To *stand* alone is the next stage; and any one who has watched the attempts of a little child to stand upright and walk will be convinced that he is moved to this by a natural instinct.[1]

It is an important epoch in a child's life when he succeeds in standing alone. Whole sets of muscles, heretofore

---

[1] Sigismund graphically describes the child's first attempts to stand in these words: " Das Kind ist selbst von seiner Verwegenheit überrascht, steht ängstlich mit weit gestellten Füssen, und lässt sich bald etwas umsanft nieder " [88].

scarcely used, are now brought into activity, and his progress is, from this time on, more all-sided and symmetrical. Hitherto his locomotion has been only in the form of creeping (which is performed in a great variety of ways, some children paddling straight ahead on all fours, like little quadrupeds, some hitching along in an indescribable manner on their haunches, and some going backwards, crab-fashion); but for the child who has learnt to stand alone, the transition to walking is, in a very literal sense, "only a step." The first conscious steps are taken very timidly, and with an evident fear of falling. But frequently the first steps are taken unconsciously. Sometimes a child who has learnt to walk, partially or wholly, reverts for a season to creeping, for no apparent reason. Children who have older brothers or sisters are likely to walk at an earlier age than others, on account of the example and assistance of these older ones. At first the feet are placed disproportionately wide apart, giving rise to a curious waddling motion; while sometimes a child *runs* instead of walking, and staggers, with the body inclined forward, and the hands stretched out as though he were afraid of falling, the feet, too, being lifted higher than is necessary. Many children seem more amiable after they have learned to walk, doubtless on account of their newly acquired ability, which not only occupies their attention, but enables them to go more readily to the objects of their desire [31].

It is perhaps scarcely necessary to call attention to the fact that a movement may be instinctive and yet not make its appearance at the very beginning of life; nor to the fact that instincts are not absolutely invariable, but are subject both to inhibition by habits and also to natural decay from desuetude.[1]

---

[1] See Professor James' chapter on Instinct, "Principles of Psychology," Vol. II.

IV. Ideational Movements.

Finally, in virtue of the aimless and will-less execution of vast numbers of movements of the nature of those already treated,— impulsive, reflexive and instinctive,— it at length comes to pass that movements are performed which are the expression of the conscious self, the index of *will* in the true and only proper sense of the word, involving a previous representation of the end sought, and (in their earlier stages) of the movements involved in attaining that end, as well as a deliberate forth-putting of the self in conscious effort towards the attainment. To such movements, and to such only, should the name of *actions* be applied. All others are only *movements*. It must not be supposed that the little child passes *per saltum* from the condition indicated in the previous sections of this chapter, to that of explicit self-conscious activity. Indeed, it would be a very false view of child-development that represented the various stages as following one another in rigid succession, with hard and fast lines showing where the one ends and the next begins. They are rather to be compared to surfaces, whose boundaries, vaguely outlined, overlap each other. There are a few impulsive movements, and very many reflex and instinctive ones, persisting to the end of life.

We shall find it convenient to follow Professor Preyer's subdivision of ideational movements into three classes. In the lowest class, we have movements of *imitation*, which, though indicating activity of will (at least in their later stages), yet depend on a model or pattern, and are never performed by the child unless he first observes their performance by others. Next, we have *expressive* movements, which, as the name indicates, are a more or less conscious expression of feelings and desires; and finally, the full-fledged *deliberative* actions.

(a) IMITATIVE MOVEMENTS. — These may be divided into two species, viz.: *Simple* imitation, in which the movement is only an approximate imitation, and no second attempt is made; and *persistent* imitation, "which marks the transition from suggestion to will, from the reactive to the voluntary consciousness." The former is exemplified in the single, isolated attempt on the child's part to reproduce a sound made by another person; the latter, in the repeated efforts of a girl of fourteen months to put a rubber on a pencil, after having seen her father do it [7], or of a boy of twelve months, to get a cord into the hole of a spool [30].

Two points should be mentioned before we proceed to record observations in this connection. First: When a child for the first time voluntarily imitates a given movement, which he has already performed involuntarily a number of times, he does it far less perfectly than when he did it without conscious imitation. "If I clear my throat, or cough purposely, without looking at the child, he often gives a little cough likewise, in a comical manner. But if I ask: 'Can you cough?' he coughs, but generally copying less accurately " [72]. Second: It must not be supposed, even when the child imitates a movement deliberately and with a clear idea of it, that he understands in every case the meaning of the movement. One child, in the tenth month, had learned to imitate the movement of beckoning, but he showed by the expression of his face and the attendant gestures, that he did not in the least comprehend the significance of the beckoning [72].

As early as the third and fourth months, according to one writer, children perform little tricks which indicate the buddings of the imitative propensity. Raw attempts at vocal imitation may be observed even in the second month, when the child makes a response to words addressed to him. This, however, is mechanical. In the third month the child will imitate looks, *i.e.*, he will look at an object which others are looking at [66]. Egger saw, in the sixth month, an instance of imitation, together with the act of recollection which it involves. Champneys says of his child: "About the thirteenth week he began to appear to attempt to join in conversation, with a variety of articulate sounds, if talking was going on in the room." Preyer observes: The first attempt at imitation occurred in the fifteenth week, the child making an attempt to purse the lips when one did it close in front of him. In the seventeenth week, the "protruding of the tip of the tongue between the lips was perfectly imitated once when done before the child's face, and the child in fact smiled directly at this strange movement, which seemed to please him" [72].

There is no point on which I find so much uniformity as this, that imitation begins during the second half of the first year. This is true of almost all children without exception, so far as I know, and extends not only to movements proper, but also to vocal imitation, as we shall see. A boy of seven months tried hard to say simple monosyllables after his mother [101]. Another is reported to have accomplished his first unmistakable imitations when seven months old, in movements of the head and lips, laughing, and the like. Crying was imitated in the ninth month, and in the tenth, imitation of all sorts was quite correctly executed, though even at the end of the first year *new* movements, and those requiring complex coördination, often failed [72]. A child of eight and a half months, having seen his mother poke the fire, afterwards crept to the hearth, seized the poker, thrust it into the ash-pan, and poked it back and forth with great glee, chuckling to himself [11]. Another child, in his tenth month, imitated whistling, and later, the motions accompanying the familiar "pat-a-cake," etc. In his eleventh month he used to hold up the newspaper, and mumble in imitation of reading [M]. Another boy, in his eleventh month, used to cough and sniff like his grandfather, and amused himself by grunting, crowing, gobbling and barking in imitation of the domestic animals and birds [C]. A little girl of this age used to reproduce with her doll some of her own experiences, such as giving it a bath, punishing it, kissing it, and singing it to sleep.

One fine morning in May I took the little boy, R., for a walk through a beautiful avenue, where the trees on each side met overhead in a mass of foliage. These trees were full of birds, busy with their nest building, and full of song. The little fellow was fairly enchanted. He could not go on. Every few steps he would stop (at the same time pulling at my hand to make me stop, too), and looking up into the trees, with his head turned on one side, would give back the bird-song in a series of warbling, trilling notes of indescribable sweetness. I very much doubt whether any adult voice, however trained, or any musical instrument, however complicated, could reproduce those wonderful inflections. The same boy, a little later, used to imitate with his voice the boys whistling in the street, giving the right pitch. Another boy, at thirteen months, brushes his hair, tries to put on his shoes and stockings, and many other similar things [C]. Indeed the whole life of the child of this age is full of imitation. Going out with the girl, F., I observed that she did almost everything I did; I brushed some dust from my coat and she immediately "brushed" her dress in like manner. It is in fact difficult fully to realize how the child of this age is watching our every movement, and learning thereby. Not only parents and teachers, but every one who comes in contact with the child, even casually and occasionally, contributes his share, whether he will or not, in the child's education. The moral of this is too obvious to require repetition.

(b) EXPRESSIVE MOVEMENTS. — These arise out of those already treated of. Impulsive, reflex, instinctive and even the simpler imitative movements, are not intentional expressions of mental states. But a movement which was at first impulsive or reflex may become the manifestation of such states. The first cry and the first puckering of the mouth (which Kussmaul noticed in children less than an hour old, when a bitter substance was brought into contact with the tongue) are only the reaction of the organism to external stimuli. But later, both the cry and the gesture fall within the control of the will, and are transformed into the purposive utterances of the conscious self. Many, perhaps most, of the expressive movements are impulsive or other movements which have been thus transformed.

The first so-called *smile*, for example (which may be observed in children less than two weeks old), is simply an impulsive movement resulting from agreeable feeling; and a reflex laugh may be elicited from a child very early by tickling the soles of his feet. In one case the first real smiles were observed from the 26th day; and in the eighth week enjoyment of music was manifested by laughing and smiling, accompanied by lively movements of the limbs, and a bright, gleaming expression of the eyes. The imitative laugh began about the ninth month [72]. Egger thinks the time when intelligence, properly speaking, appears, is marked by the advent of the laugh, which he observed for the first time after the fortieth day. Sigismund first observed a smile in the seventh week. Many children, he says, smile first in sleep; then soon after in response to the friendly looks of others. This responsive smile he believes is the first sign of consciousness of and response to sensations received from others. Many have observed the smile as early as the second and third or even the first week, but so far as I am aware, no one attributes conscious expression to the smile of a child less than a month old. Mr. Darwin believes he saw a smile of mental origin on the forty-fifth day. M. Guyau thinks the smile is reflex in its origin. Tiedemann observed a smile in the second month, and genuine laughter in the third. So also several others. The boy, C., laughed aloud when being undressed. He was then three months old. Three weeks later, when some one was reading aloud, he laughed and cooed until the reader was obliged to stop. He evidently thought the reading was intended for his special entertainment. A boy of the same age laughed aloud one day without any apparent cause [99]. The psychic development of the smile is well stated in the following words: "The smile begins when the infant first begins to be conscious of outside things; attention gradually becomes closer and more fixed; the smile at this stage is the mere stare, vacant at first, but growing steadily more intelligent and wondering in its appearance. About the third week this begins to relax very slightly into the appearance of pleasure. At this point there comes first more and more of a glow on the face — a beaming — then in a day or two a very slight relaxation of the muscles, increasing every day. This dawning smile is often very beautiful, but it is not yet a smile. It is *almost* a smile, but I am confident no one will ever know the exact day when the baby fairly and intelligently for the first time *smiles*" [100].

On *Pouting and Pursing the Lips* as an expressive movement, Preyer observes in substance: There are three sorts of pouting, differing from each other according to the cause. First, there is a protrusion of the lips, which may be observed in some children from the first hour of life, and which is purely impulsive. Secondly, the pursing of the mouth when attention is closely strained (as in learning to write or draw). This appears as early as the fifth week, and continues to the end of life in many instances. Thirdly, the pout of sullenness, which makes its appearance much later than the others, and is not due to imitation (for it occurred where there had been no opportunity for imitation), but is undoubtedly hereditary.

The *kiss*, as an expressive action, is, on the other hand, not hereditary, but acquired. Some nations do not practice it. The child has to learn it, and he is somewhat late in learning it, as observations show. Very seldom does the child understand its meaning, or give it spontaneously, until the second year of life.

The child's *cry* is at first not expressive; and when it becomes so, it varies greatly in different children. According to one observer, "Crying took place at first without any squaring of the mouth, the sound was that of 'nga' as expressed in German. It must have been produced by closing the fauces by contact of the pillars of the fauces and the soft palate, so as to send all the sound through the nose. Vowel sounds were then produced by separating the soft palate and the pillars of the fauces, and allowing the sound to come through the mouth"[15]. He goes on to say that the child seemed to cry at first for three reasons: Loneliness or fright, hunger, or pain; and these cries seemed all different in character; but he does not say when this difference became apparent. The first crying is only *squalling;* it has no expressive intonations. The transition from the meaningless *cry* to the significant *voice*, with different cries to express different mental states, has been observed as early as the second month[25], and in other cases during the third month. The little girl W., when four months old, "expressed hunger by cries that were short and shrill, following each other rapidly, and not so loud as other cries."[1]

---

[1] For further remarks on this transition from the meaningless to the significant cry see Chap. V., sec. III.

Weeping. — The new-born do not shed tears, no matter how hard they cry. At a later period they cry and weep together, and they can also cry without weeping. But to weep without crying comes much later, and is comparatively rare in childhood. One or two cases are reported of tears being shed by children two weeks old, but most of the observations point to a later date. In one case the first tears were shed at the end of the third week, in another in the fourth week, while in other cases tears were seen to flow down the face in the sixth, ninth, twelfth, fourteenth, fifteenth and sixteenth weeks respectively. Darwin's child shed tears in the twentieth week, but as early as the tenth his eyes were moist in violent crying. He thinks that children do not usually shed tears until the second, third or fourth month. From the second year onward, children weep much more easily than at an earlier period, and, later still, the inhibition both of tears and crying is a significant mark of the growing power of the will.

Nodding the head in assent, and *shaking it* in refusal, are at first entirely different from each other in mental significance. The latter is an in-born reflexive or instinctive movement, while the former is acquired. The child who has satisfied his hunger, will turn his head from side to side in refusal of further proffered nourishment when less than a week old. This movement becomes expressive almost from the first. It is generally accompanied by the partial closing of the eyes, and often by arm-movements of "warding off." Nodding in one case was not imitated until the fourteenth month, and even then very imperfectly. Even after it was finally learnt, its meaning was often confounded with that of shaking the head. The child would shake his head for "yes," and nod it for "no." In another case, both nodding and shaking the head had become expressive by the fifteenth month [68].

Other examples of expressive movements which may be observed in children at a very early age, are the following: Clasping the hands together, or waving them very quickly back and forwards, or up and down, to express eager desire for something; reaching out with uplifted hands and extended arms for the same purpose, or even sometimes clapping the hands quickly together, after the manner of an "encore;" violent straightening of the back in anger; a curious bearing, almost indescribable, showing vanity; besides several gestures expressive of affectation, and a variety of facial expressions and vocal inflections impossible to describe. "Jealousy, pride, pugnacity, covetousness, lend to the childish countenance a no less characteristic look than do generosity, obedience, ambition." All these facial expressions and bodily movements "appear in greater purity in the child, who does not dissemble, than they do in later life" [72].

(c) DELIBERATIVE MOVEMENTS. — Finally we reach that stage — not necessarily subsequent to all the others, but partially synchronous with them — in which the will rises to its proper place as "master of ceremonies," brings into subjection the impulsive and instinctive tendencies of which we have spoken, and assumes control of the child's activities. To express this truth by saying that the faculty of will has come into being, is misleading, simply because there *is* no "faculty" of will considered as a separate entity. The will *is the person* considered as active; and, instead of saying that, with the advent of what we call ideational movements, the will is born, and with that of deliberative movements it is perfected, it would be more correct to say that these movements are the first outward indications that the child is becoming the conscious master of his own activity.

In order to perform deliberative or voluntary actions in

the proper sense of the term, it is necessary that the child should have had experience of a large number of movements of the involuntary sort. For, like the man, he can *create* nothing; the most he can do, is to *combine* and *separate*, to analyze and sythesize the materials that come to his hand. Man's greatest achievements consist simply in modifying, changing, separating, combining and rearranging familiar material. So the child in all his numerous movements accomplishes nothing absolutely new; he only uses old movements, varying them it is true, in numberless ways, but really adding nothing of his own creation. Therefore the exercise of voluntary activity requires *memory* of involuntary muscular movements previously executed. For a **voluntary movement** is one which is pictured beforehand in the imagination, or, if the movement itself be not thus pictured, the end of the movement, at least, must be. But in order to *represent*, we must first *present;* or in other words, in order to imagine a movement, either in process or in product, that movement must first have been perceived; and this means that the child must have seen it performed by others, and felt it performed by himself — involuntarily — before he could perform it deliberately. So we find that deliberative movements are gradually acquired, and supervene upon a vast number of impulsive, reflexive and instinctive movements. For example, grasping with the hand is at the beginning a pure reflex, as we have seen, but gradually, after many repetitions, this movement is remembered; actual performance of the movement has led to the formation of a mental image of it, as well as a more perfect physiological adjustment favoring its performance. So that when desire, in the proper sense of the word, takes place, attention is bestowed upon the object sought and on the movement involved, and the action is deliberately performed. So we see that a strictly deliberative movement — an action —

presupposes desire, attention and memory-images. It is therefore not to be expected that we shall find *bona fide* actions in very young infants. Preyer found no movement in the first three months which could be announced with absolute certainty as a deliberative movement. Tiedemann saw the first *intended* holding of objects in the fourth month. Another child, at six months, showed a great deal of persistent effort. " He would over and over again seem to be trying to solve the problem of the hinge to his nursery door, patiently and with rivetted attention opening and shutting the door. Day after day saw him at his self-appointed task " [11]. A boy of eleven months, in striking a spoon against another object, would suddenly change it to the other hand, apparently testing whence the noise proceeded. When fourteen months old, while playing with a tin can, he put the cover on and off " not less than seventy-nine times without stopping a moment, his attention meantime strained to the utmost " [72]. Indeed the child's attention seems capable of surprising prolongation in connection with muscular movement. A little girl of nineteen months brought out her toy blocks to show me. I helped her to build houses with them. Delighted with this play, she showed a surprising persistence; and when I grew tired and wished to stop, she made me keep on longer [F]. It is by means of this incessant activity that the child develops both mentally and physically.

The ability to inhibit movements, though often difficult to observe with accuracy, seems to me one of the most certain criteria of the presence of will. To keep himself from moving is surely more difficult than to move, in a being so constitutionally restless as the average child. Children of five months [21], others of six [66], and others of seven or eight months [7], have been observed to *refrain* from reaching for an object that was much beyond their reach. The little boy R., when threatened with punishment for continued crying, is able to desist.

The development of *desire* and *attention* has perhaps been sufficiently indicated in the foregoing paragraphs. Desire in the proper sense of the word, is the primary stage in every volition; and no volition can take place without attention. The child's attention is *comparatively* weak and intermittent. He cannot attend to the unimpressive, the stimulus must be strong, must be on the motor side, and must be frequently renewed. His attention is very easy to *obtain*, but very hard to *retain*. This double fact in his nature renders him capable of education, but at the same time makes his education a gradual process, which must consist largely in the formation of right habits in him through imitation, to which, as we have seen, he is so excessively prone. M. Guyau indeed goes so far as to say that by a judicious use of the child's susceptibility to imitative suggestion, we may make of him almost what we please. And this seems indeed not far from the truth, when we consider the child's wonderful susceptibility to every passive impression, and his no less wonderful predisposition to reproduce it in his own untiring activity.

CHAPTER V.

LANGUAGE.[1]

THE profound psychogenetic significance of the language function, not only as an *index* of mind development, but also as a *factor* in that development, justifies its treatment in a separate chapter. Such separate treatment would not otherwise be justifiable, inasmuch as language does not constitute a new psychic phenomenon, or class of phenomena, differing in any essential respect from those already treated. It rather partakes of the nature of them all, and constitutes a grand product of their conjoint operation.

In order to the employment of language of any sort,[2] there must be, in the first place, *sensation*. If sounds are to be intelligently uttered, they must first be heard. The child who is born deaf, and continues in that condition, does

---

[1] This chapter first appeared as an article entitled, "The Language of Childhood," in the *American Journal of Psychology*, Vol. VI. No. 1.

[2] Although our chief attention is occupied here with the spoken word, this is by no means the only form of language. In its broadest sense, language includes every means by which thought is communicated; and therefore the gestures of the deaf-mute, and the hieroglyphic characters on Egyptian monuments, as well as the written manuscript and the printed page, are as really language as the most eloquent oral paragraphs, because they are the expression of some one's thought. As Broca says, language is "the faculty of establishing a constant relation between an idea and a sign," whatever that sign may be. All that can be said, therefore, concerning the psychological importance of the spoken word, applies equally, *mutatis mutandis*, to every other means of communication.

not learn to speak. In the second place, language presupposes *perception* and *judgment*. The sounds must not only be heard, they must be understood. A meaning must be attached to them. Otherwise they will never be given back by the child as the expression of his thought; *i.e.*, as his *language*. In the third place, it is essential to any advance beyond the merest linguistic rudiments, that abstraction and generalization take place: for the communication of thought, in its highest forms, cannot take place until there has been attained the comprehension of the general as distinguished from the particular, and of the abstract as distinguished from the concrete.[1] Finally, passing from the cognitive to the volitional aspect of mind, it is obvious that language, in its most essential characteristic — *i.e.*, as expression — belongs to the will. Every expression of thought, whether it be word or mark or gesture, is the result of an act of will, and as such may be classed among movements.

It is not, therefore, as constituting a new order of facts, different from thoughts and feelings and volitions, but rather as illustrating the development of these, and entering as a factor in that development, that language receives this separate place. We judge of the child's mental development largely by the rapidity of his progress towards a skillful manipulation of the instruments of expression.

I. Heredity vs. Education in Language.

There is no psychological problem to the solution of which a study of the infant mind may be expected to contribute more largely than this: What is hereditary, and what is

---

[1] On the other hand, thought itself cannot attain to any great degree of generality without the aid of language. Thought and language are mutually helpful, and conduce each to the development of the other.

acquired, in the sphere of language? Long before maturity is attained, such an abundance of acquired material has been added to our original store, and through constant repetition, the two have become so transformed, modified and assimilated in character, that we are no longer able to distinguish the one from the other. But from the beginning it was not so. If a child executes a gesture, or utters a sound, at an age so early as to exclude the possibility of imitation or spontaneous invention on his part, we may conclude that the sound or the gesture — or, at least the disposition to express himself in this manner — has been born with him. Here only, then, are we able to apply the logical method of difference to the solution of the problem.

It is obvious at a glance, that speech is a product of the conjoint operation of these two factors: *heredity* and *education*. If, on the one hand, we observe the initial babbling of the infant, and notice its marvelous flexibility, and the enormous variety of its intonations and inflections — and this at an age so early as to preclude observation and imitation of others, — it will be apparent that the child has come into the world already possessing a considerable portion of the equipment by which he shall in after years give expression to his feelings and thoughts. If, on the other hand, we carefully observe him during the first two years of his life, and note how the intonations, and afterwards the words, of those by whom he is surrounded, are given back by him — at first unconsciously, but afterwards with intention — and how, when conscious imitation has once set in, it plays thenceforth the preponderating *rôle*, — we shall readily believe that, without this second factor, but little progress would be made towards speech-acquirement.

It may be well to consider briefly how these two factors enter at every point in the development of language. For example, in order to speak, the child must possess first of all a sensory and motor physiological apparatus. This physiological apparatus, including the auditory structure for the reception of sounds, the inter-central and centro-motor cells and nerve tracts for the accomplishment of connection between the impression and the expression, and the organs of vocal utterance (larynx, palate, tongue, lips, teeth), is his inheritance from the past; but in the new-born child it is all imperfect, both in structure and in functioning; and its development requires the constant moulding influence of those educating agencies by which the human being is surrounded from the moment of his entrance into the world.

Again, the *disposition* to utter sounds of all sorts, and to express states of feeling by them, is undoubtedly inherited, since, from the very beginning of life, and quite independently of all example, the child constantly exercises his vocal organs.[1] But in spite of this, so inadequate is heredity alone, that the child will not learn the language of his parents, unless he be in the society of those who employ it. If brought up among savages, he will speak their language; if among wolves, he will howl.[2]

In making this statement, we do not overlook those remarkable cases in which children have invented a language of their own, quite different from that spoken around them; and persisted for some time in using the former and entirely ignoring the latter. Mr. Horatio Hale gives an account of five different cases in which this has occurred, two in the United States and three in Canada. In one case this invented vocabulary consisted of twenty-one root-forms, out

---

[1] "Le langage est en nous une faculté si naturelle, que dès la première enfance, l'exercer est un besoin et un plaisir." — *Egger*.

[2] "It is found that young birds never have the song peculiar to their species, if they have not heard it; whereas, they acquire very easily the song of almost any other bird with which they are associated." — *Alfred Russell Wallace. Natural Selection.*

of which, by combination and modification, the children developed a complete language, by which, with the aid of gesture, all their wants could be communicated; and in all the cases the invented language was sufficient for all intercourse as between the children themselves; and was persistently used until the children were finally broken of it, by being separated or sent to school. In all these cases, it is to be observed, the child did not learn the language of his parents in the absence of those who employed it. It is also to be noted that the new language was invented, not by *one* child, but by *two*. Language is *possible* in all normal children; it becomes *actual* only in the presence of a companion. But given the companion, and scarcely any limit can be set to the possibilities of development. Indeed, Mr. Hale has given us a theory of language, in which the origin of linguistic stocks is attributed to the inventiveness of children who have become separated from their tribe when very young; and in the light of such facts as those given above, the theory seems highly probable. On the other hand, that the child shall speak any specific tongue now existing, depends on his education. He does not inherit any particular tongue or dialect. Some think he will acquire his mother-tongue with greater facility than any other, yet even this may be doubted. "Speech is hereditary, but not any particular form of speech"[72]. There may be an inherited tendency to find certain sounds difficult, such as *sh* to the ancient Ephraimite, or *th* to the modern Frenchman, but this is only a tendency, and does not prevent the child from learning any language perfectly, if his education begins early enough.

Again, the careful study of the language of signs makes it quite clear that many gestures are inherited (*e.g.*, holding out the hands to express desire, which is world-wide, and is executed by children who have never seen it done), but the development of gesture into anything like a complicated system of expression, is quite dependent on the social environment. Of course this is only another way of saying that language, being the instrument for the communication of thought, is not developed in the absence of beings to whom thought can be communicated.

Thus, then, the case seems to stand with regard to the respective spheres of heredity and education in the production of language. As regards the child's present endowment and capabilities at the moment of his entrance into the world, "he is the product, the result of the generations which have preceded him; he is the visible link which connects the past with the future." [62]; but with regard to that which he is to be, and the legacy which he in his turn shall transmit to those who shall succeed him, he is very largely dependent on his physical and social environment; and all those who compose that environment, assist, whether they will or no, in his education.[1]

II. THE PHYSIOLOGICAL DEVELOPMENT.

If the question were asked, "Why does not the new-born child talk?" two answers might be given. In the first place, there is a psychological reason, viz., he has, as yet, no ideas, and has, therefore, nothing to say[86]. In the second place, there is a physiological reason, viz., his speech-apparatus is as yet so imperfectly developed that he could not express ideas if he had them.

In the same way, if the question were asked, Why does

---

[1] "La mère, au reste, ou la nourrice, ne sont ici que des institutrices en chef; car tous ceux qui entourent l'enfant au berceau qui conversent en sa présence, participent, sans s'en douter, à cette éducation fondamentale" [23].

any person ever lose the power of speech? similar answers might be given. He either loses his ideas, through some mental disorder, or he loses the power of expression through some physiological disorder. The two cases are, moreover, parallel in another sense, inasmuch as the acquirement of ideas in the one case, and their failure in the other, are closely associated with, if not indeed quite dependent upon, the presence or absence of the physiological functions.

The physiological reason, then, why the child does ot yet speak, lies in the undeveloped state of the speech-apparatus. "The lungs are not yet developed in a degree and manner sufficient for articulate speech. The expiration needs to be strong, and exactly regulated. Now, in the infant, the pectoral muscles are still developed in a very small degree; the breathing is accomplished much more through the fall of the diaphragm than through the active extension of the pectoral cavity. Hence, breathing movements are more superficial and more irregular than in later years. Artificial speech requires complete control of the breathing mechanism, which the child has not yet got. To his speech-instrument is still wanting a large number of strings, whistles and registers. The organs of speech are the lungs, air tubes, larynx and vocal cords, the mouth, with tongue, palate, lips and teeth. The lungs create the stream of air; the tone and voice are formed by the larynx; according as the vocal cords open wider or come nearer, arises the deeper or higher tone. The *form* of the tone (*i.e.*, vowel *a* or *o*, etc., consonant *b* or *f*, etc.) depends on the form of the mouth at the time. Now the larynx is still very small and undeveloped in its form, and so with the tongue, the lips, and the muscles moving them; and as for the teeth, they are still entirely wanting" [86]. The undeveloped condition of the auditory apparatus, and of the brain, have also to be considered in this connection.

On the other hand, it needs to be borne in mind that the relation between the organs of speech and speech itself is a reciprocal one. If speech depends on the organs, it is also true that the organs depend on speech, and are not developed, except by exercise. As one learns to play on the harp by playing on the harp, so the child learns to speak by speaking. The exercise of the vocal organs develops those organs, so that they become capable of higher exercise.

The *lungs* first appear, early in the embryonic stage, as a single median diverticulum from the ventral wall of the œsophagus, which soon becomes dilated towards the two sides in the form of primitive protrusions or tubercules, while the root, communicating with the œsophagus, remains single. The fœtal lungs contain no air, and lie, packed in a comparatively small compass, at the back of the thorax. They undergo very rapid and remarkable changes after birth, in consequence of the commencement of respiration. They expand so as to completely cover the pleural portions of the pericardium, their margins become more obtuse, and their whole form less compressed. The entrance of the air changes their texture so that it becomes more loose, light and spongy, and less granular; while the great quantity of blood, which, from this time on, circulates through them, greatly increases their weight, and changes their color. The proportion of their weight to that of the body becomes nearly twice as great as before, while, at the same time, their specific gravity, after the beginning of respiration, becomes very much less.

The *trachea*, or windpipe, which connects the lungs with the larynx, is in the embryo almost closed, its anterior and posterior walls being very near each other. The small space remaining is filled with mucus. With the exercise of respiration, the mucus is expelled, and the tube itself gradually becomes more distended, but its anterior wall does not for some time become convex. With the growth of the child, the cartilages which form the "ribs" of the trachea, become stronger and better able to bear their part in the forcible expiration of air which is required for speech.

The *larynx*, which is the organ most directly concerned in the production of "voice" or "tone," is an exceedingly complicated mechanism, consisting of a framework of cartilages comprising no less than nine distinct parts, connected by elastic membranes or ligaments, two of which, from their specially prominent position, are named the true vocal cords. In speaking and singing, these cartilages are moved relatively to one another by the laryngeal muscles. The larynx is situated at the upper end of the trachea, the mucus lining of the two organs being continuous. At the time of birth, this organ is very small and narrow, and continues comparatively insignificant up to the period of adolescence, when rapid and remarkable changes take place, especially in the case of the male, where it becomes much more prominent, and the *pomum adami* protrudes so to be perceptible at the throat.

The *tongue* is composed very largely of muscular fibres, running in various directions, such as the superior and inferior lingual muscles, which move the organ up and down, and the transverse fibres, by which it is moved from side to side. Besides these, we have the various glossal muscles, which, though extrinsic to the tongue itself, yet are implicated in its operations. These muscles are all more or less flabby in the fœtus and the new-born, and become strong only by nutrition and exercise. A similar remark applies to the lips; while the teeth, without which the dental and labio-dental consonants can never be properly pronounced, are at the beginning of life entirely absent, though the first steps toward their formation take place as early as the seventh week of the period of gestation[75].

The *brain* of the fœtus is comparatively deficient in convolutions, and presents a smooth, even appearance. The first of the primary fissures to appear is the fissure of Sylvius, which is visible during the third month. The other primitive sulci also begin to appear about this time, and by the end of the fifth month are well established. The secondary sulci make their appearance from the fifth or sixth month on. The first of these to be seen is the fissure of Rolando. "By the end of the seventh month, nearly all the chief features of the cerebral convolutions and sulci have appeared. The last fissures to appear are the inferior occipito-temporal, and a small furrow crossing the end of the calloso-marginal"[75]. But long after the extra-uterine life begins, the child-brain is still deficient in many of the higher processes, the association fibres being the last to develop. The convolutions are for a long time comparatively simple, and their increasing complexity as life advances stands to the exercise of the various faculties, partly in the relation of antecedent, and partly in that of consequent.

Speech, then, in the little child is a potentiality, though not an actuality. He is, as it were, in possession of the machine, but the belts have not yet been adjusted to the pulleys, nor has he yet learned to handle the instrument. The inability to speak is not, therefore, an *abnormal* state at the beginning of life, any more than the inability to write, or swim, or play the piano[72]. It is merely an *imperfect* state. But the inability to *learn* to speak is abnormal, and its cause must be sought, not in immaturity, but in abnormality of the physiological or psychological structures and processes involved. The one is an unnatural condition, into which the child has fallen; the other a natural condition, out of which he will gradually rise.

III. THE PHONETIC AND PSYCHIC DEVELOPMENT.

We shall here, first of all, give a sort of outline history of the speech-progress of the average child during the first two years, generalizing from a large number of actual observations (made by different persons on different children) and proceeding by periods of six months each; then we shall give summarized statements of a number of child-vocabularies that have been carefully compiled at different ages; and finally, we shall examine what general conclusions may be drawn from the material at hand, and set down as empirical laws, awaiting further substantiation. I say "empirical laws," because children differ so much from each other, and reliable observations are so comparatively scanty, that, for the present, general statements must be held in abeyance, or made only tentatively.

FIRST SIX MONTHS. — "In Thuringia," says Sigismund, "they call the first three months 'das dumme Vierteljahr,'" and during the second three months, according to Schultze, no advance is made on the first. It might seem, then, that in this first half-year there is nothing worthy of our attention in the matter of language. This, however, is very far from being the case, for in this period a most important apprenticeship is going on. The little child, even in the cradle, and before he is able to raise himself to a sitting posture, is receiving impressions every waking moment from the environment; is hearing the words, seeing the gestures, and noting — in a manner perhaps not purely involuntary — the intonations of those around him; and out of this material he afterwards builds up his own vocabulary. Not only so, but during this period, that peculiarly charming infantile babble (which Ploss calls "das Lallen") begins, which, though only an "awkward twittering" [66], yet contains in rudimentary form nearly all the sounds which afterwards, by combination, yield the potent instrument of speech. A wonderful variety of sounds, some of which afterwards give the child difficulty when he tries to produce them, are now produced automatically, by a purely impulsive exercise of the vocal muscles; in the same way as the child at this age performs automatically many eye-movements, which afterwards become difficult, or even impossible. M. Taine thinks that "all shades of emotion, wonder, joy, willfulness and sadness" are at this time expressed by differences of tone, equaling or even surpassing the adult [98].

The child's first act is to cry.[1] This cry has been variously interpreted. Semmig calls it "the triumphant song of everlasting life," and describes it as "heavenly music" (himmlische Musik); Kant said it was a cry of wrath, and others have spoken of it as a sorrowful wail on entering this world of sin; or as the foreboding of the pains and sorrows of life. It seems more scientific, though less poetic, to accept the explanation of the "unembarrassed naturalist," who sees in it nothing more nor less than the expression of the painfulness of the first breathing — the rush of cold air upon the lungs [88].

A more important point is the relation of this first vocal utterance to the speech that is to follow. The cry at first is merely an automatic or reflex "squall," without expressive modulation or distinctive timbre; the same cry serves to express all sorts of feelings. But very soon it becomes differentiated and assumes various shadings to express various mental states. This differentiation begins at differ-

---

[1] "Sobald das Kind zur Welt geboren ist, fängt es an gellend zu schreien" [88]. "The child is born into the world! He enters it struggling; a scream is his first utterance" [50].

ent times in different children. A girl only fifteen days old expressed her desire to be fed by a particular sort of cry [66]. In another case, the cry had ceased to be a mere squall by the end of the first month. In another, the feelings of hunger, cold, pain, joy and desire were expressed by different sounds before the end of the fifth week [74]. Others report the transition from the "cry" to the "voice" [25], involving coöperation of the mouth and tongue, at different times, but all within the first three months.

These cries are variously described. According to one, "the cry of pain is generally longer continued than the cry of fear" [49]. Another speaks of the cry of fear as "short and explosive," while hunger is expressed by a long drawn out wail [31]. Another child at two months expressed pleasure and pain by different forms of the vowel *a*. Sigismund's boy, in his sixth month, expressed pleasure by a peculiar crowing shout, accompanied by kicking and prancing.

The next step is taken when these cries and babblings assume an articulate character. The alphabetic sounds begin to be heard. Of these, the vowels usually precede the consonants; and of the vowels, *a* with its various shadings is generally the first to appear.[1] In one case the fol-

---

[1] It is necessary at this point to adopt a system of diacritical marks, as in all that follows the child's pronunciation is of great importance. We shall, therefore, adopt the following system, and shall take the liberty of changing, wherever necessary, the spelling of the recorded observations, for the sake of uniformity:

a as in *calm*.	*ē* or *ee* as in *eat, feet*, etc.	*oo* as in *food*.
ă as in *fat*.	*i* as in *pit*.	*ŏŏ* as in *foot*.
ā as in *fate*.	*ī* as in *ice*.	*u* as in *up*.
â as in *awl*.	*o* as in *pot*.	*ū* as in *use*.
ä (German *a* umlaut).	*ō* as in *old*.	*ü* (German *u* umlaut).
e as in *pet*.	*ö* (German *o* umlaut).	

Some changes will also be made in the use of consonants. For

lowing series was developed: *ä-a-u* [86]. In another, the sound of *a-a*, as an expression of joy, was heard in the tenth week [72]. According to Löbische, the vowels developed in this order: *a-e-o-u-i*. One child began with *a*, and then proceeded to *ai-ä-au-â*, while the pure sound of *ō* was late in appearing. In another case all the vowels were heard in the first five months, *ä* being the most frequently employed; and in another, the primitive *a* (of which the child's first cries largely consisted) became differentiated into the various vowel-sounds during the first month [11]. Preyer reports the use of the vowel-sounds in the following order: *uï-uo-ai-auo-ä-o-a-ö-ü-e-ä-i-u;* and Sigismund in the following: *a-ä-u-ei-o-i-ö-ü-äu-au*.

Long before the sixth month, the primitive vowels are combined with one another (as we see) and with consonants, to produce the first syllabic utterances. These first syllables are, for the most part, mechanical. In a great many of the cases under consideration, the first consonants to make their appearance are the labials, *b-p-m*, and these are almost always initial at first, and not final. The easy consonant *m*, combined in this way with the easy vowel *a*, yields the familiar combination *ma*, which, by spontaneous reduplication, becomes *mama*. In a similar manner, *papa*, *baba* (afterwards *baby*) and the like, are constructed. The labials are not always, however, the first consonantal sounds uttered. Sometimes the gutturals (*g* or *k*) precede them; and the two consonants which are usually the last to appear (viz., *r* and *l*) are used by some children quite early. In the case of the boy A., the first sounds were guttural, *gy*.

---

example, such words as *corner*, *chorus*, *coffee*, etc., will be spelled with a *k*; words like *cigar*, *centre*, *cellar*, etc., with an *s*; and in such words as *write* the silent *w* will be omitted. Other changes will be indicated as they are made.

though the earliest combination was *mam-mam*, used in crying. At five months "he dropped the throat-sounds almost entirely, and began the shrill enunciation of vowels;" and at six months he lowered his voice and began to use lip-sounds, simultaneously with the cutting of his first teeth. In another case, *m* appeared as the first consonant in the second month and was followed by *b-d-n-r*, occasionally *g* and *h*, and very rarely *k*; the first syllables were *pa-ma-ta-na* [71]. Löbische observed the consonants in this order: *m-(w)-b-p-d-t-l-n-s-r*; Sigismund in this: *b-m-n-d-s-g-w-f-ch-k-l-r-sch*; and Dr. Brown in this: *b-p-f-r-m-g-k-h-t-d-l-n* [11]. In some cases nearly all syllables have been correctly pronounced during the first half-year; while in others progress is much slower, very few syllables being certainly mastered before the ninth month.

We may sometimes observe here also the beginnings of vocal imitation. The boy A. was observed to "watch attentively the lip-movements of his attendants;" and other observers have remarked, from about the fourth month, "a curious mimicry of conversation, imitating especially the *cadences*, so that persons in the adjoining room would think conversation was going on" [72]. The same thing was observed in A. a little later.

SECOND SIX MONTHS. — Most children make a very marked advance during this period in the imitation of sounds, in the intentional use of sounds with a meaning, and in the comprehension of the meanings of words and gestures. The actual vocabulary of most children at this age is, however, exceedingly small. Many children, a year old, cannot speak a single word, while the average vocabulary does not probably exceed half a dozen words.

A new advance accompanies the rise of active hearing, and the increasing power of attention in the third three months. The child begins to keep a sort of time to music, in which he shows pleasure, and this strong excitement stimulates the production of new sounds. He delights in being carried about with a galloping rhythmic motion, and will smack his lips and make other sounds in imitation of chirping to a horse [M]. He pats his hands together in imitation of the accompanying motions in a nursery rhyme, and sometimes makes an attempt to say the words also. He shows a fondness for ringing the changes on certain syllables which he has learned, varying and reduplicating: *e.g., mama, baba, gaga, nana,* etc., and other less intelligible combinations.

He understands many words which he cannot pronounce, and he pronounces some in a mechanical way without understanding. He knows each member of the household by name, and will reach a biscuit to the person named to him. He knows the principal parts of his own body, and will point to them when asked [M]. The words *papa* and *mama,* whose surprising universality may be partly accounted for by the physiological law of case (the sound most easily produced and, therefore, earliest used, being naturally associated with those persons whose presence arouses the earliest and most vivid emotions and ideas), are by many children at this time intelligently used, though some are later in this.

Imitation usually makes rapid strides in this period. In one case gestures were imitated at eight months, and words at nine. If some one is being called, the child also calls loudly. In another case, towards the end of the child's first year, he began to imitate the sounds made by animals and inanimate objects [72]. Sigismund observed the instinct of imitation showing itself in the third quarter of the first year; the reduplication of syllables composed of a labial or dental consonant and the vowel $ä$; and the more frequent occurrence of syllables in which the vowel is initial.

Champneys' child distinctly imitated the *intonation* of the voice when any word or sentence was repeated to him several times. This has been observed also in other cases [M].

Children who are able to use a few words at this age, show by their use of them how inadequately defined is their meaning. A little girl, who had learned to say *á gá* (all gone) and *gá gá* (gegangen), applied the latter term to herself when falling down [M]. Humphreys says the child he observed was able, at this time, to name many things correctly, and to pronounce all initial consonants distinctly, except *th, t, d, c.* and *l*. Some final consonants were indistinct. Another child, at eleven months, knew what *guten tag* meant, and responded with *tata;* he also answered *adieu* with *adaa*. In this case, the first association of a sound with a concept was *ee*, which meant *wet* [72]. A boy of ten months used intelligently the words *mama, Aggie (Maggie,* this afterwards became *Waggie)* and *addie (auntie)*. At eleven months, *Waggie* was shortened to *Wag,* and *addie* to *att* [A]. Another at seven months used to wave his hand and say *tata* at parting; and one day he did this when a closet door was opened and shut again [12]. Taine's little girl at twelve months, on learning the word *bébé,* as connected with the picture of the infant Jesus, afterwards extended it, curiously enough, not to all babies, but to all pictures. Occasionally a word is invented, such as the word *mum,* reported by Mr. Darwin, which the child used with an interrogatory sound when asking for food, but also " as a substantive of wide signification." I observed a similar general use of *da,* in the case of F. In another case, the word *bo* was used to signify anything that pleased the child. The words *mama, papa* and *babē,* which had been used for some time mechanically, were dropped about the middle of this period, to be resumed five months later, " when they were applied to their proper objects" [11]. Sully observed in the beginning of this period (which he calls the *la la* period) the rise of spontaneous articulation. Combinations of syllables were suddenly hit upon, and repeated without any meaning, except as indications of baby feeling. Long *ä* indicated surprise, and "a kind of *o*, formed by sucking in the breath, indicated pleasure at some new object." In one case, a little sentence — which really consisted of two words — was uttered by a child at the close of this period. He said: "*Papa —— mama*," which meant: "Papa, take me to mama" [66].

The wide differences among children make it unsafe to venture any generalizations, except one, viz., this second half-year seems to be *par excellence* the period of the rise of imitation. Some children, however, are as far advanced at the beginning of this period as others are at its end. Perhaps it ought also to be remarked that the child who shows a great precocity in imitation, or in learning to speak, will not necessarily, on that account, turn out a more intelligent child. Imitation does not require a very high degree of mental acuteness, and the child who has been slow in this *may* in the end surpass his more precocious companion.

THIRD SIX MONTHS. — While the child is learning to walk, there is very often a standstill, or even a retrograde movement in the matter of speech. After walking is mastered, the acquisition of language goes forward again with greater facility than ever.

During this third period, marked progress is usually made in the understanding of words, and in their intelligent application, though the vocabulary is still very limited, and the pronunciation imperfect. Difficult sounds are omitted, or replaced by easier ones. Sometimes the change in one consonant has an influence on another which precedes or follows it. In longer words and combinations, only the prominent part — the accented syllable, or the final sound — is reproduced. A final *ie* is often added to words. The child says *dinnie* for *dinner*, *ninnie* for *drink*, and *beddy* for *bread*. Other imperfect pronunciations are: *apy tee* (*apple tree*), *piccy book* (*picture book*), *garny* or *nannie* (*grandma*), *pee* (*please*), *pepe* (*pencil*), *mo-a* (*more*), *hō* or *hâ* (*horse*), *Balbert* (*Gilbert*), *Tot* (*Topf*), *Ka-ka* (*Carrie*), and *Kakie* (*Katy*).

Most children at this age understand a great deal of what is said to them. Such phrases as "bring the ball;" "come on papa's knee;" "go down;" "come here;" "give me a kiss," are perfectly understood and obeyed. Parts of the child's body, as eyes, nose, ear, other ear, hand, etc., other person's eyes, ears, etc., are pointed to when named. Articles are fetched, carried and put where one commands.

Some children begin, towards the end of this period, to express themselves in short sentences, which are usually elliptical, or, as Romanes says, "sentence-words," only the most prominent word or words in the sentence being pronounced. *E.g.*, *ta* (*thank you*), *nee* (*take me on your knee*)[69]; *det off; det up; where cows George?* (*where are Uncle George's cows?*) [34]; *mo-a, mama* (*give me more, mama*); *dao* (*take me down from my chair*) [69]. Many combinations of words are made, which fall short of the dignity of sentences. *E.g.*, *mama dess, ding-a-ling*, etc. A boy of eighteen months "knows the last words of many of Mother Goose melodies, as *moon O; place O; bare, bare, bare;* putting them in at the right time, enthusiastically" [1].

Some words are invented by the child. *E.g.*, the word *tem*, which Taine's little girl spontaneously used as a sort of general demonstrative, "a sympathetic articulation, that she herself has found in harmony with all fixed and distinct intention, and which consequently is associated with her principal fixed and distinct intentions, which at present are desires to take, to have, to make others take, to look, to make others look." The same child invented the word *ham* to signify "something to eat," just as Darwin's boy used *mum* for the same purpose.

The love of reduplication shows itself very distinctly now, as indeed it has almost from the beginning; no doubt for the physiological reason that it is easier for the vocal organs to execute a movement over again, to which they are adjusted, and which they have performed once, than to adjust themselves to a new movement. Very probably the love of repetition and "jingle" which is so noticeable in children (and which, as Sigismund says, lies at the foundation of rhyme), also enters as a factor here. Numerous examples of the onomatopoetic naming of animals and things may also be observed at this time, though many of these are, no doubt, imitated from grown-up people. One or both of these tendencies may be observed in such expressions as the following: *dada, mama, papa, wawa* (*water*), *wah wah* or *ona ona* or *bow wow* (*dog*), *es es* (*yes*), *ni ni* (*nice*), *ko ko* (*chicken*), *puff* (*wind*), *quack quack* (*duck*), *golloh* or *lululu* (*all rolling objects*), *bopoo* (*bottle*), *too too* (*cars*), *tuppa tuppa tee* (*potato*), etc. The child imitates (often spontaneously) the sounds made by the dog, cat, sheep, ticking of clock, etc., while many sounds are reduplicated. The opposite process, a spontaneous curtailing of certain words, may be sometimes noticed. In one case a boy of fifteen months contracted *papa, mama* and *addie* into *pa, ma* and *att* respectively, having never heard any of these latter words [A].

Imitation is now very strong. The child attempts to repeat everything he hears; but some sounds give him difficulty, and the shifts to which he resorts in such cases are of very great interest. The boy R. used to say *nana* for *thank you*, and *dit taut* for *get caught* (in play); but the phrase *excuse me* was too much for him; he therefore used *ōhō* in its place, with a rising inflection on the second syllable. Singing is often imitated better than speech. A boy of fourteen months "gave back a little song, in the right key" [89]; and another, in the sixteenth month, knew some simple little hymns [100].

But perhaps the most interesting thing of all at this time is the gradual "clearing" of the childish concepts, as indicated by the steady circumscription of the application of names. Even yet, however, names are applied much too widely; much more experience is necessary before they acquire, in the young mind, a clear and definite connotation. (Even in mature life, most of our concepts are still very hazy and ill-defined; and it might be allowable to describe the whole process of intellectual education as a process of clarification of the concept.) It is interesting, also, to note how the principle of association enters as a factor in the determination of the application of the name. When a child calls the moon a lamp, or applies his word *bô* (*ball*) to oranges, bubbles, and other round objects; calls everything *bow wow* which bears any sort of resemblance to a dog (including the bronze dogs on the staircase, and the goat in the yard); applies his words *papa* and *mama* to all men and all women respectively; makes his word *cutie* do duty, not only for *knife*, but also for *scissors, shears, sickle*, etc. [43]; says *bâ* (*bath*) on seeing a crust dipped in tea [96]; applies *ati* (*assis*) to *chair, footstool, bench, sitting down, sit down*, etc. [25]; *peudu* (*perdu*) or *atta* (*gone or lost*) to all sorts of disappearances; — it is evident that one great striking resemblance has overshadowed all differences in the objects. Another child, who had learned the word *ōt* as a name for objects that were too warm, extended it to include, also, objects that were too cold (association by contrast). Later, on looking at a picture, he pointed to the representation of clouds and said *öt*. The clouds reminded him, no doubt, of the steam from the tea-kettle [96]. This aptitude for seizing analogies, which Taine believes to be the source of general ideas and of language, has numerous illustrations, not only in the language of the child just learning to speak, but also in the use of words by uncivilized or semi-civilized peoples.[1]

FOURTH SIX MONTHS. — During the latter half of the second year linguistic progress is usually so rapid as to render a detailed account impossible. We can only call attention, with examples, to some of the most striking features.

"By the end of the second year," says Schultze, "the normal child can make himself understood in a short sentence." His own child was able, at nineteen months, to use sentences containing subject, predicate and object. In another case, quite a complicated sentence (but very elliptical, only the nouns being uttered), was heard in the twentieth month. In the case of A., a genuine sorrow was the occasion of his first sentence. His father, of whom he was very fond, had been playing with him for some time, and finally, being called away, put him down and went out, closing the door behind him. The child gazed for a moment at the closed door, and then, throwing himself on the floor, cried out, *I want my papa*. Before this he used to express himself chiefly in elliptical sentences and sentence-words. When slightly over two years of age, he used to weave little stories of his own: *e.g., mama få wite downy toppy houf*. One day, while the dinner was waiting for his father, who was expected home on the train, the child said: *Toot-toot comy mite up tair, inny doh, uppy tapool; toot-toot make big noise*.

---

[1] See Romanes' "Mental Evolution in Man," Chap. VIII.

Another of his sentences was: *Take a badie bidy to; badie tiehd, feepy.* The boy C. uttered his first sentence in the twenty-first month: *Pees mama.* Two months earlier he had used sentence-words; e.g., *papa cacker* (*papa has fire-crackers*). In the twenty-fourth month he told quite an extensive story, in which the verbs were not expressed. Even compound sentences, and sentences containing subordinate clauses, are often mastered before the close of this period. Sometimes verbal inflections appear; e.g., *naughty baby klide* (*cried*) [60]. Another day the same child said *comed* for *came*, thus unconsciously rebuking the inconsistent English language. Interrogative sentences appeared in another case; e.g., *where's pussy?* and negation was expressed by an affirmative sentence, with an emphatic *no* tacked on at the end, exactly as the deaf-mutes do. Many of these primitive sentences are in the imperative mood, and many are still "sentence-words." Most children talk a great deal, and gesticulate profusely, at this time. Their expressions are concrete, and abstract words are avoided as far as possible. A little boy, on seeing the picture of a half-grown lad, spoke of it as a *little baby man* [A]. Anything that has rhyme or rhythm is most easily picked up. A little nephew of my own was able, at this age, to sing a large number of little songs and hymns, giving the melody quite correctly. Another boy, at twenty-one months, on hearing the milkman's bell in the morning, used to say: *Mik man mik cow, crump horn, toss dog, kiss maid all floru;* or peeping through the fence at the cows, would sing: *Moo cow, moo cow, how-de-do cow.* [1]

The child's progress is marked here by his gradual mastery of the personal and possessive pronouns. These are peculiarly difficult for the average child, and, according to Egger, are not usually attained until near the close of the second year; according to others, much later still (thirtieth month, according to Lindner). Previous to mastering the *I*, the child calls himself by his proper name, or by the name *baby*, as he may have been taught. When *I* first appears, it is frequently employed,—quite consistently from the child's point of view,—not in the first person, but in the second; *i.e.*, he calls others *I* and himself *you*. One child used the word *I* correctly as early as the nineteenth month, but often exchanged it for her proper name [86]. Another, in the twentieth month, still called himself by his proper name, but, a month later, said *me* for the first time [96]. Another spoke of *me* as a personality in her twenty-second month [100]. Another, at two years, often used the word *my*, meaning *your*; *e.g.*, *Let me get up on my lap* [38]. Another, at the same age, still speaks of himself as *baby* in ordinary converse, but in great desire says, *I want it*, and in great fear says, *I afraid*.

In some cases, almost all the sounds are mastered by the end of the second year, but from the observations at hand, this may be considered the exception. Most children still have difficulty with certain sounds. Some of these difficulties are seen in the following: *apou* (*apple*), *zhatis* (*there it is*), *es* (*yes*), *yleg* (*egg*; note difficulty with initial vowel), *oken* (*open*), *tash* (*mustache*), *sh'ad* (*thread*), *dam* (*gum*), *t'al* (*shawl*), *appervator* (*elevator*), *nobella* (*umbrella*), *bannicars* (*banisters*), *aw yi* (*all right*), *setto* (*cellar*), *pato* (*potato*), *it da* (*sit there*). One observer reports a special difficulty with *s, z, d, g, k, l, n, q, r* and *t* [13]. Another says that at nineteen months, the sounds *s, sh, ch* and *j* were generally indistinct; while *w, r* and *f* were formed, but not well developed. On the other hand nasal *g* appeared, *o* was mastered, *l, p* and *t* as *final* consonants began to be used, and *k* became a favorite sound, used in many words. Sibilants were more at command when final than when initial, while short *ă* was just beginning to be formed. In the twenty-second month the sounds of *ch, j* and *th* were still imperfect, the hard sound of *th* being replaced by *s* and the soft sound by *z*. A month later, *r* was still generally replaced by *l*; when *s* came before another consonant, one or the other was dropped, and *k* was sometimes confused with *p* or *t* [69]. In another case, the double consonant *sp* made its first appearance at the end of the second year [88].

There are still many examples of the inadequate limitation of the concept. In one case the word *poor*, which was learned as an expression of pity, was applied on occasion of any sort of loss or damage whatsoever, and was even used in speaking of a crooked pin. *Dam* (*gum*), with which toys were mended, became a universal remedy for all things broken or disabled; and afterwards, when the child acquired the word *sh'ad* (*thread*), broken things were divided into two classes, viz., those that were to be mended with *dam*, and those that were to be mended with *sh'ad* [69]. *Behwys*, in another case, was at first the name for all small fruits, but afterwards became restricted, yielding a portion of its territory to *gape* (*grape*) [A]. Another little boy extended his word *gee-gee* (*horse*) to a drawing of an ostrich, and a bronze figure of a stork; and his word *apoo* (*apple*) to a patch of reddish-brown color on the mantelpiece [96]. The boy C. applied the word *bōke* (*broke*) to a torn pocket-handkerchief; and R. extended his word *dō* (*door*) to everything that stopped up an opening or prevented an exit, including the cork of a bottle, and the little table that fastened him in his high chair.

Healthy children of two years of age will usually attempt all sorts of sounds in imitation of others, and will practice on new and difficult combinations with great perseverance, sometimes carrying the word through several stages of transition, until it finally assumes the perfect form. The boy A. first heard the word *pussy* when seventeen months old; he at once undertook to say it, but called it at first *pooheh*, then *poofie*, then *poopoohie*, then *poofee*, until finally, after much persevering practice, he was able to say *pussy*, when he seemed to be satisfied, and discontinued its use, except when pussy was in sight. Schultze gives, among others, the following examples: The German word *wasser* passed through these stages, — *wawaff* — *fafaff* — *waffwaff* — *wasse* — *wasser*; the word *grosmama* was first *ōmama*, and then *dōsmama*, before assuming its final form. The strength of the reduplicating tendency, and the influence of the initial consonant on the remainder of the word, is seen in the following imitations: *wawa* (*Mary*), *dudu* (*Julia*), *ih ih* (*little*), *ba ba* (*blanket*), *fafa* (*faster*), *mama* (*master*), *papa* (*pasture*), *nana* (*naughty*) [36].[1]

---

[1] I cannot forbear quoting the following from Sigismund in this connection. A child of twenty-one months attempted to repeat, line by line, a piece of poetry after another person. The first line in each pair represents the pronunciation of the adult, the second the imitation of the child:

Guter Mond, du gehst so stille,
Tute Bohnd, du tehz so tinne.

Durch die Abendwolken hin,
Duch die Aten-honten in.

Gehst so traurig, und ich fühle,
Tehz so tautech, und ich büne.

Dass ich ohne Ruhe bin,
Dass ich one Ule bin.

Guter Mond, du darfst es wissen,
Tute Bohnd, du atz es bitten.

Weil du so verschwiegen bist,
Bein do so bieten bitz.

Warum meine Thränen fliessen,
Amum meine Tänen bieten.

Und mein Herz so traurig ist,
Und mein Aetz so atich iz.

VOCABULARIES. — I have taken the trouble to collect, for purposes of comparison, a number of vocabularies of children, which have been recorded by careful and competent observers, with as much completeness and accuracy as possible. I will now give these in summarized form, so as to show the relative frequency of the various sounds as initial, and also the relative frequency of the various parts of speech. In order the more accurately to show the sounds actually made by the child, I have been obliged to use an alphabet differing somewhat from the ordinary English alphabet. The following changes are made: *c* is dropped out altogether, such words as *corner, candy,* etc., being classed under *k*; words like *centre, cigar,* etc., under *s*; and words like *chain, cheese, chair,* etc., forming a new series under *ch*. Words like *George, gentleman,* etc., are classed under *j* instead of *g*; words like *Philip* under *f*; words like *knife, knee,* etc., under *n*; and words like *wrap, write,* etc., under *r*. Other new letters besides *ch* are *sh* and *th*. In short, it is sought to classify the child's words according to his pronunciation, and not according to the English alphabet. If he says *tátie* for *potato,* the word is classed under *t*. I am convinced that this is the only way to obtain reliable and valuable results.

I. A child of nine months is reported as speaking "nine words plainly." The words are not given [100].

II. A boy at twelve months has "four words of his own" [100].

III. A child of twelve months uses ten words with meaning. Six of these are nouns, two adjectives and two verbs [66]. The initial sounds are *m* (three times), *p* (four times), *n, a* and *k* (each once).

IV. A child of one year used eight words, seven of which were nouns, and one an adverb. The initial sounds are *b* (four times), *m, p, d* and *u* (one each) [T].

V. The boy R. had at command about twenty words, thirteen of which were nouns, and four or five interjectional words. For initial sound *b* was perferred, then *p* and *t*.

VI. Another child is reported, at fifteen months, as having "syllables, but no words" [100].

VII. A girl of seventeen months is reported as using thirty-five words, twenty-two of which are nouns, four verbs, two adjectives, four adverbs and three interjections. The initial sounds are *d* (eight times), *s* (four), *m*, *b* and *ch* (three each), *p*, *t*, *k*, *a* and *y* (two each), *i*, *j*, *u*, *o* (one each) [L].

VIII. A girl of twenty-two months uses twenty-eight words, distributed as follows: Nouns sixteen, verbs three, adjectives three, adverbs and interjections five. The initial sounds are *b* (six times), *d* (five), *m* (four), *p* (three), *g*, *h* and *k* (two each), *e*, *i*, *u* and *o* (one each) [G].

IX. A girl at two years employs thirty-six words, distributed as follows: Nouns sixteen, adjectives four, pronouns three, verbs seven, adverbs three, interjections three [G]. Initial sounds are, *p* (five times), *m*, *b* and *w* (each four times), *g*, *k* and *h* (each three times), *d*, *i*, *u* and *r* (each twice), *a* and *o* (each once).

X. Summary of the vocabulary of a little boy in Washington, D.C., aged nineteen months (K).

	Nouns.	Adj.	Pron.	Verb.	Adv.	Prep.	Conj.	Interj.	Total.
A	2	1		1					4
B	10	2		2				1	15
Ch	1								1
D	4	2		1	1				8
E									
F	1	1		2					4
G	4	3		2	2			1	10
H	3			5					8
I	1								1
J									
K	8			3				2	13
L				2					2
M	5	1		1					7
N	2	1			1				4
O	2	1		1		1			5
P	6	1		5					12
Q									
R	2								2
S	13	1		9				2	25
Sh									
T	7	1		1					9
Th	2								2
U					1				1
V	4	1		1					6
W	1	3		1					5
Z									
Total	78	19		36	4	1		6	144

XI. Vocabulary of a girl of twenty-one months; the daughter of an Andover professor (K).

	Nouns.	Adj.	Pron.	Verb.	Adv.	Prep.	Conj.	Interj.	Total.
A	4							1	5
B	22	1	1		1				25
Ch	5								5
D	7	1		1	1				10
E	1								1
F	6								6
G	6			1					7
H	9			1					10
I	2								2
J									
K	17	2		1					20
L	3								3
M	9	1							10
N	7								7
O	2					1	1		
P	11			1					12
Q									
R	5			4					9
S	14			4					18
Sh									
T	11				1				12
Th	2								2
U			1		1				2
V	2								2
W	5	1		1					7
Z									
Total	148	6	2	14	5	2			177

XII. Vocabulary of a girl of twenty-two months in Worcester, Mass. (F).

	Nouns.	Adj.	Pron.	Verb.	Adv.	Prep.	Conj.	Interj.	Total.
A									
B	6							3	9
Ch									
D	3			2			1		6
E				1			1		2
F	1			1			1		3
G									
H	1	1		1					3
I				1					1
J									
K	1			1					2
L									
M	3	1	1	2			1		8
N	3				1				4
O				1	1		1		3
P	3	1		1			1		6
Q									
R									
S									
Sh	2	1		1					3
T	9			1				1	11
Th									
U						1			1
V									
W	3								3
Z	2			1					3
Total	37	4	1	9	8			10	69

XIII. Vocabulary of a girl in Brooklyn, N.Y., in her twenty-third month (S).

	Nouns.	Adj.	Pron.	Verb.	Adv.	Prep.	Conj.	Interj.	Total.
A	1	2					1		4
B	22			2				1	25
Ch	2								2
D	8				2				10
E	4	1							5
F	1								1
G	1			2					3
H	2			1					3
I	3	1		1					5
J	2								2
K	10			1					11
L	2			1					3
M	8	1	2	1					12
N	3				1				4
O	1		1				1		3
P	11	1		4					16
Q									
R	1	1							2
S	9	1		1					11
Sh									
T	3	2		1					6
Th	1				1				2
U				1					1
V									
W	1	1		1	1				4
Z				1					1
Total	95	12	3	16	7		1	2	136

XIV. Vocabulary of a little country girl of American parentage, in her twenty-fourth month (K).

	Nouns.	Adj.	Pron.	Verb.	Adv.	Prep.	Conj.	Interj.	Total.
A	4	2		1	1				8
B	20	1		4				1	26
Ch	5								5
D	7	1		2	1				11
E	1			1					2
F	10	1		2					13
G	7	1		6				1	15
H	8	4		6					18
I	2		1	1		1			5
J	1								1
K	24	3		5					32
L	1	1							2
M	9	1	3						13
N	4	1		1	1				7
O	3	1			1	1		1	7
P	15	1		3					19
Q	1								1
R	5			4					9
S	17	2		5					24
Sh	3			1					4
T	13			4					17
Th	2	1		1	1				5
U	1		1			2			4
V									
W	5	1		5	1				12
Z			2		1				3
Total	168	22	7	52	7	4		3	263

XV. Vocabulary of a Vermont boy, twenty-four months old (C).

	Nouns.	Adj.	Pron.	Verb.	Adv.	Prep.	Conj.	Interj.	Total.
A	1								1
B	13	1		4				4	22
Ch	7							1	8
D	7	1		1	2				11
E	2								2
F	2								2
G	3								3
H	1	1		1				1	4
I	1								1
J	2								2
K	11	1		3				1	16
L	4								4
M	8	1	1						10
N	3				1				4
O									
P	12			1				1	14
Q	1							1	2
R									
S	7			7					14
Sh									
T	5			2				1	8
Th									
U									
V									
W	5	1		2	1			2	11
Z									
Total	95	6	1	21	4			12	139

XVI. Vocabulary of a Boston boy, slightly over two years old (A).

	Nouns.	Adj.	Pron.	Verb.	Adv.	Prep.	Conj.	Interj.	Total.
A	7	1			1			1	10
B	28	2		4					34
Ch	3								3
D	14	2		6	2				24
E	3	2		1	1				7
F	9			6					15
G	7	1		2					10
H	10	2		1	2	1			16
I	1	1	2			1			5
J	2								2
K	22	2		2					26
L									
M	15	1	1	2					19
N	6	4		1	2				13
O	1			1	1	2		1	6
P	16	3		3				1	23
Q									
R	2			1					3
S	14	3		7					24
Sh	3								3
T	10	3		5					18
Th									
U					1	1			2
V	1								1
W	13	2		4	1				20
Z		1							1
Total	187	30	3	46	11	5		3	285

XVII. Vocabulary of a girl in her twenty-fifth month (K).

	Nouns.	Adj.	Pron.	Verb.	Adv.	Prep.	Conj.	Interj.	Total.
A	4	1							5
B	24			5					29
Ch	4								4
D	13	2		3	1				19
E	3								3
F	9								9
G	6	1		2					9
H	19								19
I	2	1	1	1		1			6
J	3								3
K	23	2		3					28
L	5	1		1					7
M	9		1						10
N	7	1			1				9
O	2								2
P	24	1		3					28
Q	2				1				3
R	7	1		1					9
S	5	2		5					12
Sh	3								3
T	18			3		2			23
Th									
U					1				1
V									
W	4	2		1	1				8
Z			1						1
Total	196	15	3	28	5	3			250

144

Summarizing these vocabularies, we find some interesting facts bearing on language-growth, both on the physiological and on the psychological side.

For example, with regard to the relative frequency of the various parts of speech, the following table is instructive. Of the five thousand four hundred words comprising these vocabularies [1]

60 per cent are	nouns.
20 " " "	verbs.
9 " " "	adjectives.
5 " " "	adverbs.
2 " " "	pronouns.
2 " " "	prepositions.
1.7 " " "	interjections.
0.3 " " "	conjunctions.
100.0	

Of the nouns, less than one per cent are abstract. Nearly all are names of persons or familiar objects. The majority, in the earlier months, seem to be used almost with the force of proper nouns, as Schultheiss has also observed. The adjectives are mostly those of size, temperature, cleanliness and its opposite, and similar familiar notions. This table also corroborates Sigismund's observation that the conjunction is especially difficult. Another interesting point is the comparison of the above table with a similar table, showing the relative frequency of the various parts of speech in ordinary adult language. Professor Kirkpatrick says that of the words in the English language,

---

[1] In all the calculations that follow, I have taken the liberty to *include*, along with my own vocabularies, those of Professor Holden, and Professor Humphreys, which I have re-arranged phonetically for the purpose.

60 per cent are nouns.
11 " " " verbs.
22 " " " adjectives.
5.5 " " " adverbs.

An important consideration is involved here. If we look only at the first of these two tables, and consider the child's words by themselves, it will *seem* that the nouns have greatly the advantage over the other parts of speech. But such a conclusion obviously cannot be drawn, unless a comparison of the child's vocabulary with that of the adult justifies us in so doing. In order to show that the child learns nouns more easily than verbs, we must be able to show that the number of his nouns bears a larger proportion to the number of nouns he will use as an adult, than the number of his verbs bears to the number of verbs he will use in adult life. To represent the matter symbolically.

Let n = the proportion of nouns in the child's vocabulary.
And N = " " " " " " man's "
Let v = " " " verbs " " child's "
And V = " " " " " " man's "

Then, if the child learns nouns more easily than verbs, the proportion of n to N will be greater than that of v to V. But on comparing the two tables, the very opposite is found to be the case.

$$\text{For } \frac{n}{N} = \frac{60}{60} = 1$$

$$\text{But } \frac{v}{V} = \frac{20}{11} = 1.81 +$$

In other words, the child of two years has made nearly twice as much progress in learning to use *verbs* as in learning to use *nouns* ; according to my tables of child-language and Professor Kirkpatrick's table of adult-language.[1] A comparison of the adjectives and adverbs in the two tables justifies a similar conclusion in favor of the adverb. To my mind, this fact — which, so far as I know, has been hitherto overlooked by all writers on child-language — possesses great value for philology and pedagogy as well as for psychology. In the first place it supports the view that the acquisition of language in the individual and in the race proceeds by similar stages and along similar lines. Max Müller says that the primitive Sanscrit roots of the Indo-Germanic languages all represent *actions* and not *objects* ; that in the race the earliest ideas to assume such strength and vividness as to break out beyond the limits of gesture and clothe themselves in words are ideas of movement, activity. We have found, from examination of the vocabularies of these twenty-five children, that the ideas which are of greatest importance in the infant mind, and so clothe themselves most frequently (relatively), in words, are the ideas of *actions* and not *objects*, of *doing* instead of *being*. The child learns to use *action-words* (verbs) more readily than *object-words* (nouns) ; and words descriptive of actions (adverbs) more readily than words descriptive of objects (adjectives).[2]

---

[1] This statement is still further confirmed by a vocabulary received since the publication of the first edition. It is the vocabulary of a five-year-old boy in Minneapolis. Of the sixteen hundred words spoken by this boy, 19 per cent were verbs and only 53 per cent nouns.

[2] Professor Kirkpatrick, in a private note, suggests that, since his tables of adult language are taken from the dictionary, they very likely do not represent truly the vocabulary of the average adult. It appears that, in " Robinson Crusoe," the proportion of nouns to verbs is not 60 to 11, but 45 to 24. If " Robinson Crusoe " represents the average adult vocabulary, then the conclusions stated in the text will need

In the second place this fact confirms the Froebelian principle, on which child-education is coming more and more to be based, viz., that education proceeds most naturally (and, therefore, most easily and rapidly) along the line of motor activity.[1] The child should not be so much the receptacle of instruction, as the agent of investigation. Let him *do things*, and by doing he will most readily learn. He should not be *passive*, but *active* in his own education. The kindergarten is the modern incarnation of this idea, but the idea itself is as old as Aristotle, who says, "We learn an art by *doing* that which we wish to do when we have learned it: we become builders by building, and harpers by harping. And so by doing just acts we become just, and by doing acts of temperance and courage we become temperate and courageous."[2]

Turning now to the consideration of these vocabularies from the standpoint of *ease or difficulty of pronunciation* of the various simple sounds, we find some instructive data here also. The following table shows the relative frequency of the various sounds *as initial*. In this calculation no heed is paid to the English spelling of the words, but only to the sounds actually uttered by the child, as already pointed out. Of the five thousand four hundred words

---

revision. I imagine, however, that in a book so full of *action* as "Robinson Crusoe," the verb element would be unusually strong.

[1] My colleague, Professor van der Smissen, gives me the very interesting observation, that his little girl, who is just learning to talk, uses many sentences in which the verbs are not *spoken* at all, but *acted*, all the other words in the sentence being spoken. *E.g.*, "Willie whipped the pussy," would be expressed by the words, "Willie . . . pussy," accompanied by a lively slapping movement of one hand upon the other.

[2] "Eth. Nic.," Book. II. Chap. 1. par. 4.

11	per cent begin with the sound of						b.
10.3	"	"	"	"	"	"	s.
9	"	"	"	"	"	"	k.
8	"	"	"	"	"	"	p.
6.1	"	"	"	"	"	"	h.
6	"	"	"	"	"	"	d.
6	"	"	"	"	"	"	m.
6	"	"	"	"	"	"	t.
5.2	"	"	"	"	"	"	w.
4	"	"	"	"	"	"	f.
4	"	"	"	"	"	"	n.
3.2	"	"	"	"	"	"	g.
3.1	"	"	"	"	"	"	l.
3	"	"	"	"	"	"	a.
3	"	"	"	"	"	"	r.
2	"	"	"	"	"	"	i.
2	"	"	"	"	"	"	sh.
1.3	"	"	"	"	"	"	th.
1.2	"	"	"	"	"	"	e.
1.1	"	"	"	"	"	"	o.
1	"	"	"	"	"	"	ch.
1	"	"	"	"	"	"	j.
1	"	"	"	"	"	"	y.
0.8	"	"	"	"	"	"	u.
0.5	"	"	"	"	"	"	v.
0.2	"	"	"	"	"	"	q.

A glance at this table shows how prominent a place the explosive consonants occupy as initial sounds in child-language.[1] The vowels, on the contrary, though undoubtedly

---

[1] The vocabulary of the five-year-old Minneapolis boy, spoken of in a previous footnote, conforms, in the main, to this order. The five sounds he used most frequently as initial are s, p, b, k, f, in the order named.

the earliest sounds to be used in most cases, are very infrequent *as initial*, not only absolutely but relatively. In the English dictionaries the vowel *a* occupies fourth place as initial letter [43]; in my tables it occupies fourteenth place; while the other vowels stand still lower. The reason of this is not far to seek. It is simply a case of the operation of the law of physiological ease; as any one may verify by pronouncing, in succession, the following syllables: *ap, pa, ab, ba, ak, ka, am, ma, ad, da;* and observing how much more easily those syllables are pronounced in which the consonant leads and the vowel follows.

Another interesting feature of this table is the high place occupied by the guttural *k* as initial sound. It stands above *p* and *m*, and next to *s* and *b*. This fact does not bear out the theory propounded by several writers on child-language, that those sounds are selected by the child for earliest acquirement whose pronunciation involves those portions of the vocal apparatus which are most easily seen, such as the lips. According to this theory, not only the labial *p*, but the sounds *d, m, f, sh, th,* etc., ought to stand high in the list, because the movements involved in their pronunciation are plainly visible: while the guttural *k*, whose movements are absolutely out of sight, should stand very low. The contrary is the case; *k* stands third in the list of initial sounds, while *th*, whose movements are exceedingly obvious to sight, occupies the eighteenth place. This seems to prove that the child does *not* learn to utter sounds by watching the mouths of those who utter them in his presence; and this opinion is confirmed by the observation of Schultze, that the child does not usually look at the *mouth*, but at the *eyes* of the person speaking to him. On the other hand there seems no sufficient ground for the statement that the law of least effort is overturned by this frequency of the sound of *k*. This guttural sound is, for most children, no more difficult than the labials. Often it is one of the very earliest sounds employed. I know one child with whom it is more frequently used than even *b*. In short, so far as my observations go, I have no hesitation in saying that the child's earliest vocal utterances are not acquired by imitation at all, either of sound or of movement, but that they are purely impulsive in their character. They are simply the result of the overflow of motor energy, which we have seen so prominent in other departments of the child's life; and they proceed at first along the lines of least resistance.

In the following tables I have given the results of a careful examination of seven hundred instances of mispronunciation which I have found in the above vocabularies. The first table shows the various sounds in the order of the number of times they are misused, as well as the ways in which they are misused; the second and third tables enter into more detail.

In the following table the first column gives the sound misused; the second shows the number of times it is replaced by another sound; the third shows how often it is dropped, without being replaced; and the fourth shows how often it is brought into a word to which it does not belong (not as a substitute for some other sound, but as a pure interpolation, for no apparent reason).

Sound Misused.	Re- placed.	Dropped.	Inter- polated.	Total.	Sound Misused.	Re- placed.	Dropped.	Inter- polated.	Total.
R	51	87	4	142	W	7	5	3	15
L	35	70		105	Ch	13			13
S	25	34	1	60	Y	1	10	1	12
G	25	6		31	V	8	2		10
T	13	17	1	31	E	2	5		7
Sh	26	4		30	H	2	5		7
K	20	8		28	J	5			5
Th (hard)	21	2		23	P	4	1		5
					A		4		4
F	15	4	1	20	M	4			4
D	5	12	2	19	Wh	3			3
Th (soft)	14	4		18	O	3			3
					B	3			3
Ng	15			15	Z	1		1	2
N	7	7	1	15	Q	1			1

The following table shows the relative frequency of *replacement* of the sounds when initial, medial, and final, and also (in the case of the consonants) when occurring as one member of a double consonant (*e.g.*, as *r* in *cream*). It also gives the relative frequency of the substituted sounds:

Sound Replaced.	When Initial.	When Medial.	When Final.	When Double.	Replaced by	Times.	Examples.
					w	29	kweem (cream).
					l	6	tommollä (tomorrow).
R	21	21	9	4	y	3	all yīte (all right).
					e	8	tumblie (tumbler).
					v	1	voom (room).
					t	1	tautech (traurig).
					m	1	pipe (ripe).
					p	1	Kaka (Carrie).
					k	1	

LANGUAGE.

Sound Replaced.	When Initial.	When Medial.	When Final.	When Double.	Replaced by.	Times.	Examples.	
L	8	8	19	3	e	9	minnie	(milk).
					w	7	tabie	(table).
					u	7	singu	(shingle).
					n	4	setta	(celery).
					t	3	bampe	(lampe).
					b	2	degen	(legen).
					d	2	apoo	(apple).
					oo	1		
Sh	17	2	7		s	19	fis	(fish).
					h	4	hōogar	(sugar).
					b	1	tooz	(shoes).
					t	1		
					n	1		
S	18	4	3	6	t	8	tweet	(sweet).
					h	8	hlate	(slate).
					f	3	poofee	(pussy).
					b	3	dīde	(side).
					d	3		
G	19	5	1	4	d	17	dass	(glass).
					k	2	hookoo	(sugar).
					t	2	toss	(gross).
					b	1	bavy	(gravy).
					w	1	detting	(getting).
					j	1		
					n	1		
Th (hard)	11	3	7	5	f	10	free	(three).
					t	4	mous	(mouth).
					s	3	tank	(thank).
					p	1	harf	(hearth).
					d	1	nuppin	(nothing).
					n	1		
					r	1		
K	11	7	2	7	t	15	bastet	(basket).
					s	2	sun	(come).
					g	2	untle	(uncle).
					d	1	tanny	(candy).
F	7	4	4	2	p	6	nup	(enough).
					s	5	buttersy	(butterfly).
					k	2	kork	(fork).
					t	2	ōt	(off).

Sound.	When Initial.	When Medial.	When Final.	When Double.	Replaced by.	Times.	Examples.
Ng		5	10	1	n e a	12 2 1	finner (finger). tockies (stockings). lockatair (rocking chair).
Th (soft)	11	3			d m	13 1	altogedder (altogether). dare (there).
T			7		e k w g p	6 4 1 1 1	dockie (doctor). bankie (blanket). jackie (jacket). coak (coat). wawer (water).
Ch	9	2	2	1	s t sh	7 4 2	sair (chair). tillens (children). shick (chick).
V	1	5	2		b f d	5 2 1	gib (give). shufer (shovel). Dadie (David).
N		1	6		e m l	4 2 1	buttie (button). pim (pin). lemolade (lemonade).
W	6	1			v l	6 1	go vay (go away). lalla (water).
D	1	4			n t k	2 2 1	towntownt (down town). vinner (window). kankie (candy).
J	4	1			d g	4 1	demidon (demijohn). Gekkie (Jessie).
P	3		1	1	b t	2 2	bee (please). patie (paper).
M	2	2			k n w	2 1 1	hankie (hammer). Waggie (Maggie).
Wh	3				f h	2 1	feel (wheel). haiah (where)

Sound Replaced.	When Initial.	When Medial.	When Final.	When Double.	Replaced by.	Times.	Examples.	
O			3		a	2	winna	(window).
					ĕ	1		
B	1	2			d	2	badie	(baby).
					m	1	Milly	(Billy).
E			2		ă	1	vera	(very).
					oo	1	cookoo	(cookie)
H	1	1			t	1	torns	(horns).
					l	1	lä lo	(la haut).
Y		1			ē	1	bēwo	(bureau).
Z		1			d	1	Döderfeen	(Josephine).
Q		1			k	1	skeeze	(squeeze).

The following table gives similar information with regard to the *dropping* of difficult sounds:

Sound Dropped.	When Initial.	When Medial.	When Final.	When Double.	Examples.	
R	2	61	24	50	each	(reach).
					apicot	(apricot).
					dotta	(daughter).
					baselet	(bracelet).
L	10	37	23	39	etta be	(let me be).
					peeze	(please).
					fā	(fall).
					buttafy	(butterfly).
S	27	4	3	30	poon	(spoon).
					Bottie	(Boston).
					gā	(gas).
					tūbewie	(strawberry).

Sound Dropped.	When Initial.	When Medial.	When Final.	When Double.	Examples.	
T		9	8	8	dissance	(distance).
					bonny	(bonnet).
					sottin	(stocking).
D	1	5	6	12	sanny	(sandy).
					gamma	(grandma).
					bines	(blinds).
Y		6	4		ard	(yard).
					panna	(piano).
K	4	2	2	2	opf	(kopf).
					basset	(basket).
					bŏŏ	(book).
N	1		6	1	pi	(pin).
					burr	(burn).
G	6			1	atten	(garten).
W	5				ont	(want).
					ŏŏdn't	(wouldn't).
F	3		2		nuff	(enough).
					koff	(coffee).
H	5				eah	(here).
Sh	4				litta	(schlitten).
F		3	1	2	sātie pin	(safety pin).
					nātanoon	(afternoon).
Th (soft)	3	1			ăt	(that).
					ōber air	(over there).
A	4				fāde	(afraid).
					nudda	(another).
Th (hard)			2		bă	(bath).
					mao	(mouth).
V	1		1		ammum	(warum).
					Duttie	(Gustave).
P	1				tātie	(potato).
Z			1		nō	(nose).

A word of caution is perhaps necessary here. These tables do not show accurately the order of difficulty of the various sounds, inasmuch as they indicate the misuse of the sounds, not relatively to the number of *correct* pronunciations of each sound, but only relatively to the total number of mispronunciations. For example, in the first table *q* seems an easier sound than *b*, because it is only misused once, while *b* is misused three times. But if we remember that in the vocabularies *b* occurs fifty-five times as often as *q*, the case is entirely altered. Considered in this way, the order of difficulty, according to my observations, is approximately the following: *r, l, th, v, sh, y, g, ch, s, j, e, f, t, n, q, d, k, o, w, a, h, m, p, b.* The most difficult sound is *r* and the easiest *b*.

It will be observed also that, according to these tables, mispronunciation is very frequent in the case of double consonants, and most frequent of all in those combinations which belong to what Mr. Pitman calls the *pl* and *pr* series. Such words as *cream, bracelet,* and *fly* are almost always mutilated; sometimes *r* and *l* are replaced by *w* or some other sound; sometimes they are omitted altogether.

Another thing to be observed is that the choice of a substitute for a difficult sound is often determined by the prominent consonant in the preceding or succeeding syllable. This leads to a reduplication of the easier sound in preference to the use of the more difficult one. The child says *cawkee* for *coffee, kork* for *fork,* or *la lo* for *la haut.* The number of these reduplications is very large, and the device is adopted also in the case of difficult vowels; *e. g., Deedie* occurs for *Edie,* and *Dida* for *Ida.*

Another significant thing is the frequency with which the sound of *ĕ* is used as a substitute for difficult sounds, both vowel and consonantal, especially at the end of a word. The child says *ittie* for *little, finnie* for *finger,* and *ninnie* for *drink.*

In addition to the mispronunciations tabulated above, I find a large number of miscellaneous mispronunciations difficult to classify, such as the following: *medniss* for *medicine*, *Mangie fag* for *American flag*, *skoogie* for *excuse me*, *killie* for *tickle*, *pā-tā-soo* for *patent leather shoes*, etc.

If we seek now to discover some principle underlying the development of child-speech from the psychic point of view, we shall find, I believe, that principle of *transformation*, which we have already observed so frequently elsewhere, operating in this sphere also. The earliest utterances of the new-born have little or no psychic significance. As expressions of his thought, they have none at all. But by slow degrees these primitive utterances, modified, increased and combined, are associated with ideas, which are also modified, increased and combined, until finally the instrument of language is completely under control, and becomes the adequate medium for the expression of thought.

Not only may we make this statement in this general way, but it seems possible to trace, with approximate minuteness, the progress of a sound upward, from the earliest unexpressive condition to the highest, latest, most expressive state, and to indicate the principal stages on the way. These stages appear to be the same as those through which movements pass, viz., the *impulsive*, the *reflex*, the *instinctive*, and the *ideational*. The first sounds uttered by the child are simply the spontaneous will-less, idea-less manifestation of native motor energy. They do not require a sensory, but only a motor process, and that motor process is automatic. The same overflowing energy, the same muscle-instinct, which impels the child to grasp with the hands, to kick with the feet, etc., impels him also to the exercise of his lips, tongue, larynx and lungs [46]. This is the *impulsive* stage. Then we find him uttering sounds in response to certain sensations. He sees a bright light, hears a peculiar sound, feels a soft, warm touch, and these sensations call forth certain sounds. These sounds are still only babblings, not involving the coöperation of will, but they do involve sensory as well as motor processes. The reflex arc, in its simplest form, is complete. Here imitation takes its rise. This is the *reflexive* stage. In the next place we can detect certain sounds which are expressive of the child's needs, and though still uttered probably without conscious intention, yet have a purpose and an end, viz., the satisfaction of those needs. The cry, which was at first monotonous and expressionless, now becomes differentiated to express various states of feeling, hunger, pain, weariness, etc. Here we have the *instinctive* stage. Finally the will takes full possession of the apparatus of speech, the child utters his words with conscious intention; imitation of sounds, from being passive and unconscious, becomes active and conscious; and words are joined together to give expression to ideas of constantly increasing complexity. Here we have reached the *ideational* or deliberative stage.

As an example of the transformation of a single sound through all these successive stages, let us take that sound which is, in the majority of cases, the first articulation, the syllable *ma*. At first this is pure spontaneity. The child lies contentedly in his cradle, motor energy overflows, the lips move, gently opening and closing, while the breath is expired, and this sound is produced, *mamamama*. As yet it has no meaning; it is a purely automatic utterance. But by and by the same sound is called forth by certain sensations, one of which is very probably the sight of the mother, or of some other person. The word as yet has no definite meaning, but is merely a sort of vague demonstrative ejaculation, a pure reflex. Later it becomes the expression of certain bodily needs and conditions, and now the hungry child utters this sound as the expression of the need of his natural nourishment. By this means, the word becomes firmly associated with the mother, first probably with the breast only [67], but afterwards with her person in general, and so the final step in the transition is made, and the word *mama* now passes out of the semi-conscious, instinctive stage into the ideational. It becomes firmly associated with the mother, and with her only, it is used with a conscious purpose of communicating to her the child's wishes and ideas, and, finally, in her absence, it is used in such a way as to show that her image is firmly stamped on his mind, and retained in his memory. In later life, more abstract and complex applications of this word are gradually mastered; but we have followed it far enough in its development for our present purpose. This word was chosen because it probably exemplifies better than any other the principle which we desired to illustrate, being associated with those feelings which arise earliest, last longest, and take the deepest hold upon the human soul; but almost any primitive utterance of infancy could be employed to exemplify, in a less complete manner, the principle enunciated.

UNPUBLISHED SOURCES OF INFORMATION.

A. Observations on a little Boston boy, made and recorded by Miss Sara E. Wiltse.

B. Observations made by Professor J. M. Baldwin, of the University of Princeton.

C. A little Vermont boy, whose mother, a graduate of Smith College, made a very careful record of his mental development.

D. Vocabulary kindly sent me by Professor H. H. Donaldson, of the University of Chicago.

E. Observations made by a student of Wellesley College.

F. A little girl in Worcester, Mass., whom I observed for some time, and from whose parents I received some valuable notes.

G. Two little girls in Springfield, Mass. Observations made by their mother.

K. Observations kindly sent me by Professor E. A. Kirkpatrick, of Winona, Minnesota.

L. A girl in North Carolina, aged seventeen months. Notes taken by her mother.

M. Observations made by Professor and Mrs. J. F. McCurdy, of the University of Toronto.

R. A strong, healthy Canadian boy, whom I observed during a large part of his second year.

S. Notes on a little girl in Brooklyn, N.Y., sent me by her father.

T. A little boy in Boston. Vocabulary recorded by his mother.

W. A little girl in Worcester, whose development was recorded by her mother.

Y. References to the lectures of the late Professor G. P. Young, on Philosophy and Psychology, delivered in the University of Toronto, but as yet unpublished.

PUBLISHED SOURCES OF INFORMATION.

1. ALLEN, MARGARET A. (MRS.). "Notes on the Development of a Child's Language." *Mother's Nursery Guide*, February, 1893.
2. ALLEN, MARGARET A. (MRS.). "A Mother's Journal." In *Babyhood*, March, 1885.
3. ALLEN, GRANT. "The Color-Sense: Its Origin and Development." London: Trübner & Co., 1879.
4. BALDWIN, J. M. "Handbook of Psychology: Senses and Intellect."
5. BALDWIN, J. M. "Mental Development in the Child and the Race." Vol. I. (in preparation).
6. BALDWIN, J. M. "Origin of Right and Left-handedness." *Science*, October 31, 1890.
7. BALDWIN, J. M. "Suggestion in Infancy." *Science*, February 27, 1891. "Bashfulness and its Meaning." *Educational Review* (N.Y.), December, 1894. "How to observe Children's Imitation." *Century Magazine*, December, 1894.
8. BATEMAN, F., M.D. "Aphasia, and the Localization of the Faculty of Speech." London, 1890.
9. BELL, A. G. "The Progress made in Teaching Deaf Children to Read Lips and Talk, in the United States and Canada." *Science*, August 26, 1892.
10. BINET, A. "Perceptions d'Enfants." In *Revue Philosophique*, December, 1890.
11. BROWN, ELIZABETH STOW, M.D. "The Baby's Mind: Studies in Infant Psychology." Read before the New York Academy of Anthropology, April, 1889. Published in *Babyhood*, July-November, 1890.
12. CANFIELD, W. B., M.D. "The Development of Speech in Infants." In *Babyhood*, May, 1887.

13. CHAILLE, S. E., M.D. "Infants, their Chronological Progress." *New Orleans Medical and Surgical Journal*, June, 1887.
14. CHAMBERLAIN, A. F. "Notes on Indian Child-Language." In *American Anthropologist*, July, 1890.
15. CHAMPNEYS. "Notes on an Infant." In *Mind*, Vol. VI., p. 104.
16. CHILDREN OF ALL NATIONS. A collection of papers by different writers. London: Cassell & Co.
17. CHRISMAN. O. "Child Study, a New Department of Education." *Forum*, February, 1894.
18. CLARUS, A. "Ueber Aphasie bei Kindern." Leipzig, 1874.
19. COMPAYRÉ, G. "Evolution Intellectuelle et Morale de l'Enfant." Paris, 1893.
20. CZERNY, A. "Beobactungen über den Schlaf im Kindesalter unter physiologischen Vernhältnissen." In *Jahrb. für Kinderk.* N. F., XXXIII. S. 1.
21. DARWIN, CHARLES. "Biographical Sketch of an Infant." *Mind*, Vol. II., p. 285.
22. DARWIN, CHARLES. "Expression of the Emotions." New York, 1873.
23. DEGERANDO, M. "De l'Education des Sourds-Muets de Naissance." Tome Premier. Paris, 1827.
23 (a). DEWEY, J. "The Psychology of Infant Language." *Psychological Review*, January, 1894.
24. DUPANLOUP, MGR. (Bishop of Orleans). "The Child." Translated, Kate Anderson. Boston, 1875.
25. EGGER. "Sur le Developpment de l'Intelligence et du Langage." Paris, 1887.
26. FAUST, B. C. "Die Perioden des Menschlichen Lebens." Berlin, 1794.
27. FEILING, H. "Das Dasein vor der Geburt." Stuttgart, 1887.
28. FRÖBEL, F. "Mutter-und Kose-Lieder." Translated, Frances and Emily Lord. London, 1890.
29. GARBINI, ADRIANO. "Evoluzione della voce nella infanzia." Verona, 1892.
30. GEIGER, L. "Ursprung und Entwickelung der Menschlichen Sprache und Vernunft." Two volumes. Stuttgart, 1872.
31. GENZMER, A. "Untersuchungen über die Sinneswahrnehmungen des Neugeborenen Menschen." Halle, 1882.
32. GOLTZ, B. "Buch der Kindheit." Berlin, 1847.
33. GUYAU, M. "Education et Heredité." Paris, 1890. *Bibliothèque de Philosophie Contemporaine*.

34. HALE, HORATIO. "The Origin of Languages and the Antiquity of Speaking Man." Reprinted from *Proceedings of American Association for the Advancement of Science*. Vol. XXXV.
35. HALL. G. S. "Contents of Children's Minds on Entering School." *Pedagogical Seminary*, June, 1891.
36. HALL, G. S. "Notes on the Study of Infants." *Pedagogical Seminary*, June, 1891.
37. HALL, G. S. "Child-Study the Basis of Exact Education." *Forum*, December, 1893.
38. HOLDEN, E. S. "On the Vocabularies of Children under Two Years of Age." *Trans. Am. Phil. Assoc'n*, 1877, p. 58 et seq.
39. HUMPHREYS, W. "A Contribution to Infantile Linguistics." *Trans. Am. Phil. Assoc'n*, 1880, p. 5.
40. JAMES, W. "Psychology" (Briefer Course). American Scientific Series. New York, 1892.
41. JASTROW, J. "Problems of Comparative Psychology." *Pop. Sci. Mo.*, November, 1892.
42. KEBER, A. "Zur Philosophie der Kindersprache." Leipzig, 1890.
43. KIRKPATRICK, E. A. "How Children Learn to Talk — a Study in the Development of Language — Children's Vocabularies." *Science*, September 25, 1891.
44. KITCHEN, DR. J. M. W. "Infantile Grief." *Babyhood* for June, 1892.
45. KRONER, T. "Ueber die Sinnesempfindungen der Neugeborenen." Breslau, 1882.
46. KUSSMAUL, A. "Die Störungen der Sprache." Leipzig, 1877.
47. KUSSMAUL, A. "Untersuchungen über das Seelenleben des Neugeborenen Menschen." Tübingen, 1884.
48. LAMSON, M. S. "Life and Education of Laura D. Bridgman." Boston. 1878.
49. LEONARD, W. E. "Elementary Training of Infants." In *Babyhood*. March, 1890.
50. LIEBER, F. "The Vocal Sounds of Laura Bridgman." *Smithsonian Contributions*, 1851. Vol. II., p. 6.
51. LÖBISCHE. "Die Seele des Kindes." Wien, 1851.
52. LEYS, J. "The Brain and its Functions." International Scientific Series. New York, 1882.
53. MACHADO Y. ALVAREZ — of Seville. "Titin: A Study of Child-Language." In *Trans. Philol. Soc.* (London), 1885-87, pp. 68-74.

54. MALACHOWSKI, E. "Versuch einer Darstellung unserer heutigen Kentnisse in der Lehre von der Aphasie." Leipzig, 1888. Sammlung klinischer Vorträge, No. 324.
55. MALLERY, G. "Sign Language among the North American Indians." *Report of Bureau of Ethnology*, 1879–80.
56. MARENHOLTZ-BÜLOW, BARONESS. "The Child, and Child-nature." Translated, Alice M. Christie. London, 1890.
57. MARWEDEL, E. "Conscious Motherhood." Boston, 1889.
58. MOSSO, A. "La Peur: Étude psycho-physiologique." Paris, 1886.
59. MÜLLER, F. "Grundriss der Sprachwissenschaft." Band I. Abtheilung I. Vienna, 1876.
60. NEWELL, W. W. "Games and Songs of American Children." New York: Harper's, 1884.
61. NOBLE, EDMUND. "Child Speech, and the Law of Mispronunciation." *Education*, September and October, 1888.
62. OUROUSSOV, PRINCESS MARY. "Education from the Cradle." Translated, Mrs. E. Fielding. London, 1890.
63. PEREZ, B. "L'Art et la Poesie chez l'Enfant." Paris: F. Alcan, 1888. *Bibliothèque de Philosophie Contemporaine*.
64. PEREZ, B. "Le Caractère de l'Enfant a l'Homme." Paris, 1892. *Bibliothèque de Phil. Contemp.*
65. PEREZ, B. "Education Morale dès le Berceau." Paris, 1888.
66. PEREZ, B. "The First Three Years of Childhood." Translated by Alice M. Christie. London, 1889.
67. PLOSS, H. "Das Kind, in Brauch und Sitte der Völker." Two volumes. Leipzig, 1884.
68. PLOSS, H. "Das Kleine Kind." Berlin, 1881.
69. POLLOCK, F. "An Infant's Progress in Language." *Mind*, Vol. III., p. 392.
70. POLLOCK, F. "F. Schultze, Die Sprache des Kindes." *Mind*, Vol. VI., p. 436.
71. POTTER, S. O. L., M.D. "Speech and its Defects." Philadelphia, 1882.
72. PREYER, W. "The Mind of the Child." Two volumes. Translated, H. W. Brown. New York, 1889.
73. PREYER, W. "Mental Development of the Child." Trans. W. Brown. International Education Series. New York.
  , W. "Physiologie des Embryo." Leipzig, 1885.

74. PREYER, W. "Psychogenesis." *Jour. Spec. Phil.*, April, 1881.
75. QUAIN'S "Anatomy," Vol. II. London, 1882.
76. QUEYRAT, FR. "L'Imagination et ses Variétés chez l'Enfant." *Bibliothèque de Philosophie Contemporaine.* Paris, 1893.
77. RADESTOCK. "Habit and its Importance in Education." Boston, 1886.
78. RAEHLMANN, E. "Physiologisch-psychologische Studien über die Entwickelung der Gesichts-wahrnehmungen bei Kindern und bei operierten Blindgeborenen." In the *Zeitschrift für Psychologie und Physiologie der Sinnesorgane*, Vol. II. (1891), pp. 53-96.
79. RIBOT, TH. "Heredity." New York, 1889.
80. RIBOT, TH. "Diseases of Memory." New York, 1880. Humboldt Library, No. 46.
81. ROMANES, G. J. "Mental Evolution in Man." New York, 1889.
82. ROSS, J. "On Aphasia." London, 1887.
82 (a). ROYCE, JOSIAH. "The Imitative Functions." *Century Magazine*, May, 1894.
83. SANZ DEL RIO. "Psicologia del Nino." In *Boletin de la Institucion libre de Ensenanza*, 1893, pp. 17-19 (No. 383), (and a former article in No. 372).
84. SEMMIG, H. "Das Kind: Tagebuch eines Vaters." Leipzig, 1876.
85. SCHULTHEISS, W. K. "Das Kind, in der Entwickelungszeit des Geistes." Nuremburg, 1862.
86. SCHULTZE, F. "Die Sprache des Kindes." Leipzig, 1880.
87. SHINN, MILLICENT W. "Notes on the Development of a Child." University of California Studies. Part I., 1893; Part II., 1894. Berkeley, California.
88. SIGISMUND, B. "Kind und Welt." Braunschweig, 1856.
89. SIKORSKI. "Du Développment du Langage chez les Enfants." Paris: *Archiv. de Neurol.*, VI., 1884, p. 319, etc.
90. SIKORSKI. "L'Evolution Psychic de l'Enfant." *Rev. Phil.*, March and May, 1885.
91. SOZINSKEY, T. S. "Mental Aspects of Infantile Unfoldment." *Med. and Surg. Reporter*, Philadelphia, 1882, XXVI., p. 309.
92. STEINTHAL, H. "Der Ursprung der Sprache im Zusammenhange mit den letzten Fragen alles Wissens." Berlin, 1888. Fourth Ed.

93. STEVENSON, A. "Jealousy in an Infant." *Science*. October 28, 1892.
94. STEVENSON, A. "The Speech of Children." *Science*. March 3, 1893.
95. SULLY, JAS. "Babies and Science." *Cornhill Magazine*. May, 1881.
96. SULLY, JAS. "Baby Linguistics." In *Eng. Ill. Mag.*, November, 1884.
97. SULLY, JAS. "Outlines of Psychology." New York, 1885.
98. TAINE, H. "Acquisition of Language by Children." *Mind*. Vol. II., p. 252.
99. TAINE, H. "De l'Intelligence." London, 1871.
00. TALBOT, MRS. E. "Papers on Infant Development." Published by the education department of the American Social Science Association. Boston, 1882.
01. TIEDEMANN. "Record of Infant Life." Translated. Perez Syracuse.
02. TYLOR, E. B. "Early History of Mankind." New York, 1878.
03. VIERORDT, H. "Das Gehen des Menschen." Tübingen, 1881.
04. WARNER, F. "The Children: How to Study them." London, 1887.
05. WOLFE, H. K. "On the Color-Vocabulary of Children." *Nebraska University Studies*, July, 1890, pp. 205-234.

INDEX.

Abstraction, 78.
Action, 68, 91, 102.
Affectation, 110.
Affection, 55.
Altruism, 57.
Analogy, 82.
Anger, 58, 110.
Association, 69, 70, 71, 78, 134.
Astonishment, 50.
Attention, 65, 91, 92, 111, 113.

Beautiful, feeling of, 54.
Beckoning, 103.
Blinking, 7.
Brain, 123.

Color, 14.
Concept, 76, 134, 138.
Consonants, 149.
Co-ordination, 6.
Coughing, 95, 103.
Creeping, 100.
Crowing, 93.
Cry, xii., 60, 95, 104, 106, 108, 125, 159.
Curiosity, 51.

Deaf-mutes, 76.
Denotation, 79.
Desire, 91, 92, 110, 111, 113, 118.
Differentiation, 62.
Doll, 105.
Dramatic instinct, 55.
Dreams, 46, 73.

Ear, 20, 30, 117.
Eye, 2, 6, 7, 29, 96.

Fixation, 8.
Fœtus, xi., 1, 2, 20, 27, 35, 37, 39, 41, 44, 93, 95, 123.

Generalization, 78.
Gesture, 118.
Gutturals, 150.

Habit, 94.
Head, holding, 100, shaking, 109, nodding, 109.
Hearing, 23, 24, 45, 64.
Heredity, 26, 50, 55, 61, 92, 94, 115.
Hiccough, 95.
Homesickness, 56.
Humor, 53.
Hunger and thirst, 39, 40.

"I," 85, 86, 87, 137.
Idiots, 26.
Illusions, 64.
Imagination, 57, 72, 73.
Imitation, 62, 73, 102, 103, 104, 105, 109, 128, 129, 131, 133, 138, 139, 151, 159.
Impatience, 48.
Inhibition, 95, 97, 109, 112.
Instinct, 41, 55, 58, 85, 92, 98, 117.
Invention, 117, 132.

Jealousy, 58.
Judgment, 80.

Kiss, 55, 56, 108.

Laugh, 53, 95, 106.
Light, 1, 5.

Lips, 28, 122.
Localization, 62.

Mama, 127, 159.
" Me," 87, 157.
Mirror-image, 64, 71, 84.
Mispronunciations, 148, 151.
Movement, 6, 7, 38, 39, 41, 68, 83, 91, 92, 93, 94, 96, 102, 106, 110, 111, 118.
Mouth movements, 98, 100.
Muscular feeling, 41.
Music, 24, 53, 106, 128.

Nostrils, 29, 30.

Pain, 5, 39, 40, 97.
Papa, 73, 127.
Parts of speech, 136, 145, etc.
Perspective, 12.
Pictures, 80.
Play, 51, 58.
Pouting, 107.
Property instinct, 58, 85.
Purpose, notion of, 79.

Reasoning, 80.
Recept, 77.
Recognition, 55.
Reduplication, 133, 139, 157.
Reflexes, 5, 94, 95, 96, 97, 159.
Religious instinct, 50.
Respiration, 95.
Rhyming, 71.
Rhythm, 129, 136.
Right-handedness, etc., 98.

Selfishness, 58.

Sentences, 132, 135.
Sibilants, 137.
Singing, 133.
Sitting, 100.
Smile, 53, 55, 93, 106.
Sneezing, 95.
Space, 60, 62.
Standing, 100.
Starting, 96.
Stretching, 93.
Sucking, 38, 39, 63, 81, 99.
Suggestion, 57, 89, 113.
Summation of stimuli, 68.
Surprise, 49.
Swallowing, 95.
Syllables, 127.
Sympathy, 57.

Teeth, 98, 122.
Temperature, 37, 38.
Thumb, contraposition, 98.
Tongue, 28, 100, 122.
Transformation, xii., 43, 51, 90, 94, 106, 158.

Vanity, 54, 85, 110.
Visual interpretation, 18, 19.
Vocal apparatus, 121, 122.
Vowels, 126.

Walking, 100.
Weeping, 109.
Whistling, 104.
Will, 92, 102, 110, 112.
Words, 131.

Yawning, 93.

www.ingramcontent.com/pod-product-compliance
Lightning Source LLC
Chambersburg PA
CBHW031442160426
43195CB00010BB/821